CONTENTS

Introduction

A Legal and management
01	General responsibilities	13
02	Accident reporting and recording	25

B Health and welfare
03	Health and welfare	31
04	First aid and emergency procedures	41
05	Personal protective equipment	45
06	Asbestos	55
07	Dust and fumes (Respiratory hazards)	61
08	Noise and vibration	69
09	Hazardous substances	77
10	Manual handling	83

C General safety
11	Safety signs	89
12	Fire prevention and control	95
13	Electrical safety	101
14	Work equipment and hand-held tools	107
15	Mobile work equipment	115
16	Lifting operations and equipment	121

CONTENTS

D High risk activities

17	Working at height	129
18	Excavations	145
19	Underground and overhead services	151
20	Confined spaces	157

E Environment

21	Environmental awareness and waste control	165

F Specialist activities

22	Demolition	177
23	Highway works	185
24	Specialist work at height	197
25	Lifts and escalators	211
26	Tunnelling	215
27	Plumbing (JIB)	225
28	Heating, ventilation, air conditioning and refrigeration (HVACR)	229

Further information

Training record	248

CONTENTS
Introduction

Overview	2
Foreword	2
New for the 2020 edition	3
Acknowledgements	4
Work-related injuries and ill health statistics for the construction industry	4
Occupational health and safety	5
Working Well Together in construction	6
Setting out	6
How to use GE707	8
Use of icons	8
Additional content	8
Glossary	9
Further supporting information from CITB	12
Feedback	12

INTRODUCTION

Overview

The purpose of this *Construction health and safety awareness* (GE707) publication is to explain the basic health, safety and environmental steps that you, your employer and the site or project that you are working on should be taking.

The content gives easy to understand information and practical guidance. Each chapter starts with a summary of what sites and employers should do for you, as well as what they expect from you.

GE707 is the official supporting document for the CITB Site Safety Plus one-day *Health and safety awareness course*.

 The decision to leave the European Union does cause uncertainty, not only in relation to health, safety and environmental legislation and regulation. Withdrawal from the EU will not, on its own, have an impact on the Health and Safety at Work etc. Act, but Brexit may result in the long-term re-examination of UK Regulations which were implemented to align with EU Directives. At the time of publication references to legislation and regulations are correct. CITB does, however, strongly urge you to remain alert to possible future change in this area.

 This publication contains public sector information published by the Health and Safety Executive (HSE) and licensed under the Open Government Licence.

Foreword

If you are new to construction, welcome. This exciting and varied industry can offer you a long and rewarding career. It is important to remember, though, that construction is still a dangerous industry and that we all have a part to play in staying safe. Your employer, working with CITB and other training organisations, must make sure you have the right skills, knowledge, training and experience for the job that you are doing. Inexperienced workers need more supervision, but attending a health and safety awareness course is a good start.

Your employer must make sure you have the right skills, knowledge, training and experience for the job you are doing

INTRODUCTION

Both experienced and new workers have a duty of care for the health, safety and wellbeing of themselves and others working on site, and both need the right knowledge to do that. This *Construction health and safety awareness* publication will help to explain basic health, safety and environmental information, highlighting potential hazards on site and giving you advice on how to keep safe. It covers areas such as your general responsibilities on site, causes of work-related ill health and how to prevent them, working safely with tools and equipment, and protecting the environment.

Accidents can happen when people don't follow a safe system of work (for example, when rushing to get a job completed) and some case studies within this book highlight how easily things can go wrong.

CITB is committed to giving you, and everyone who steps onto site, the right skills and knowledge for you to do your job safely, and without damaging your health or the health of those around you. We all want to make construction sites safe places where no-one is harmed. Working together, looking out for each other, we can raise health and safety standards and make our industry even better.

Revd. Eur Ing Kevin Fear BSc (Hons), CEng, MICE, CMIHT, CMIOSH, Hon FaPS
Health and Safety Strategy Lead
CITB

New for the 2020 edition

All sections have been reviewed and updated, where necessary, to ensure the information reflects current legislation and good practice. The following section-specific updates have also been made.

- Section A includes a new case study that highlights how a lack of training led to a young construction worker suffering life-changing injuries.

- In Section B the health and welfare chapter has new information about the requirements for mental ill health and first aid. The section also includes information on the requirements for training people in mental health first aid.

- In Section C additional information highlights the importance of, and legal requirement for, lifting equipment to be subject to thorough examination. For lifting this should take place at least every 12 months, and every six months for equipment used for lifting people.

- Section E includes information advising that Local Authorities (LAs) or water authorities must be informed, and a permit obtained, before the disposal of any hazardous fluids into the drainage system.

INTRODUCTION

Acknowledgements

CITB wishes to acknowledge the assistance offered by the following organisations in the preparation of this edition of GE707.

- JSP Ltd.
- Mount Anvil.

Work-related injuries and ill health statistics for the construction industry

- The construction sector is a major employer accounting for around 7% of the UK workforce.
- Construction includes three broad industry groups.
    1. Civil engineering – covering general construction for civil engineering works, including road and railway construction, and utility projects.
    2. The construction of buildings – covering general construction of buildings, including new work, repair, additions and alterations.
    3. Specialised construction activities – covering trades that are usually specialised in one aspect common to different structures (for example, demolition, electrical and plumbing installation, joinery installation, plastering, painting and glazing).
- The UK construction industry is made up of about 275,000 construction businesses, of which 90% employ fewer than 10 workers.
- Approximately 2.65 million people are employed in the UK construction industry. It covers activities including housing, utilities, repair and maintenance, refurbishment, demolition, roofing, shopfitting, mechanical and electrical, plumbing and highways maintenance.
- On average 35 construction workers are killed each year in work-related accidents.
- The biggest cause of deaths (around half) is falls from height, with an average of seven people dying each year as a result of falling through fragile roofs.
- The most common over seven-day injuries are due to manual handling or lifting accidents, followed by slips, trips and falls on the same level, falls from height and being struck by an object.
- The construction industry has the largest burden of occupational cancer. It accounts for over 40% of occupational cancer deaths and cancer registrations each year in Great Britain.
- The most significant carcinogen is past exposure to asbestos, followed by silica dust, solar radiation, and coal, tars and pitches.

INTRODUCTION

- Approximately 5,000 people die each year because of past exposure to asbestos fibres.
- Work-related respiratory disease covers a range of illnesses that are caused or made worse by breathing in hazardous substances (such as construction dust) that damage the lungs.
- Silica dust is the biggest risk to construction workers after asbestos. Prolonged exposure to respirable crystalline silica can cause lung cancer and other serious respiratory diseases.
- Vibration white finger, carpal tunnel syndrome, noise-induced hearing loss and dermatitis are the most common non-lung diseases suffered by those in the construction industry.
- Around 0.4 million working days are lost every year because of stress-related absence.

 Construction workers (just like you) could die because of work-related ill health, or as a result of an accident, if control measures are not followed.

Occupational health and safety

Occupational health deals with all aspects of health and safety in the workplace. The main focus is on the prevention of hazards.

 Occupational health **is the promotion and maintenance of the highest degree of physical, mental and social well-being of workers in all occupations. This means keeping workers as healthy as possible, by removing or preventing the hazards and risks than can cause harm to people at work.**

Safety **is freedom from physical harm. (The condition of being protected from, or unlikely to cause, danger, risk or injury.)**

Some of the work activities, equipment, materials and chemicals used in the workplace have risks that may lead to accidents, musculoskeletal diseases (long-term spine or back problems), respiratory diseases (breathing problems, from breathing in dust or fumes), hearing loss, circulatory (blood) diseases, cancers, stress-related disorders, and many other conditions. To make sure the health of workers is not affected, employers have to protect people from possible hazards at work.

When someone is injured on site the effects are usually immediate or obvious (such as a cut). However, some of the conditions mentioned above may cause harm inside the body for many years before the effects are known or felt. In some cases it can be up to 60 years before someone is diagnosed with a disease that has no cure, caused by exposure to a hazard when working in construction many years before (for example, breathing in harmful dust or fumes).

INTRODUCTION

Working Well Together in construction

The Working Well Together (WWT) campaign is an industry-led initiative that helps support micro and small businesses to improve their health and safety performance. WWT has become the most successful health and safety initiative within the construction industry. The campaign undertakes a variety of work, including health and safety awareness days, designer awareness days, breakfast and evening events, roadshows and regional WWT groups.

Working Well Together campaign posters

The following are aims of the WWT initiative.

- Improve health and safety knowledge and good practice and seek to continuously improve health, safety and welfare performance in the construction industry.
- Provide practical advice and assistance to the construction industry on the provision and maintenance of healthy and safe working environments.
- Encourage co-operation between members in relation to sharing knowledge of health and safety matters.
- Promote training and learning in health and safety.
- Provide free or low-cost information, advice and training to employers and workers in the construction industry.
- Stage events where at least half the people who attend will be from small businesses (15 or fewer employees) and 20% from micro businesses (five or fewer employees).

 To find out how the WWT campaign can help you and your company visit the website: wwt.uk.com

Setting out

Construction is an exciting industry. It is constantly changing as projects move on and jobs get done. As a result of this, a building site is one of the most dangerous environments to work in. But accidents and ill health can be avoided if everyone on site works together.

INTRODUCTION

A free online film, *Setting out*, explains what you and everyone on site must do to stay healthy and safe at work.

This short film is essential viewing for everyone involved in the construction industry. The content of the film is summarised here.

Part 1: What you should expect from the construction industry

Your site and your employer should be doing all they can to keep you and your colleagues safe.

Before any work begins the site management team will have been planning and preparing the site for your arrival. It is their job to ensure that you can do your job safely and efficiently.

Five things the site you are working on must do are listed below.

- Know when you are on site (signing in and out).
- Give you a site induction.
- Give you site specific information.
- Consult with you on matters which affect your health, safety and welfare.
- Keep you up-to-date and informed.

Part 2: What the industry expects of you

Once the work begins, it is up to everyone to take responsibility for carrying out the plan safely.

This means you should follow the rules and guidelines as well as being alert to all the changes on site.

Five things you must do are listed below.

- Respect and follow the site rules.
- Safely prepare each task.
- Complete each task responsibly.
- Know you can and should stop if you think anything is unsafe.
- Keep learning.

Every day the work we do improves the world around us. It is time for us to work together to build an industry that puts its people first. By working together we can build a better industry that respects those who work in it.

INTRODUCTION

How to use GE707

GE707 follows the standard structure that is used across all core CITB publications.

Section A: Legal and management
Section B: Health and welfare
Section C: General safety
Section D: High risk activities
Section E: Environment
Section F: Specialist activities

Each chapter begins with a summary list of what your site and employer should do, together with a list that explains what you should do for your site and employer.

Use of icons

A set of icons emphasises important points within the text and also directs readers to further information. The icons are explained below.

 Website/further info

 Example

 Question

 Ideas

 Notes

 Important

 Good practice

 Poor practice

 Caution

 Consultation

 Guidance

 Case study

 Quote

 Definition

 Interactive checklists and forms

 Video

Additional content

 This additional content (AC) icon is used in our publications to direct you to complementary content such as videos, interactive scenarios and weblinks.

To access this extra content use the following steps to navigate the structure.

The example provided is for the *Setting out* film, referenced on page 7.

INTRODUCTION

Step 1	Open CITB's companion website	www.citb.co.uk/ge707
Step 2	Open the relevant section for the content required	Supporting information
Step 3	Select the relevant link	Watch the *Setting out* film
Step 4	Access the additional content	Video opens on YouTube

Where can I find additional content in this publication?

The table below identifies the pages in this publication where the AC icon appears and the information that can be accessed via CITB's companion website.

Page	Content
7	Watch the *Setting out* film
35	Watch a needlestick injuries toolbox talk
50	Watch an eye protection toolbox talk
57	Watch Simon's story – living with an asbestos-related disease
73	Access the HSE noise exposure demonstration recordings
135	Watch the fragile roofs film

Glossary

Many words and terms that you will hear on a construction site are explained in the main part of this book. The list below includes some more terms that you might come across.

ACoP. Approved Code of Practice.

Acute. An immediate health effect (for example, burns to the skin from acids or chemicals).

Adhesive. A substance used for sticking things together.

Abrasive wheel machine. A machine, such as a bench-mounted grinder or a disc-cutter, which is used for cutting or grinding materials.

Allergy. A damaging reaction of the body caused by contact with a particular substance.

Asbestos. A naturally occurring, heat-resistant substance that was once used extensively in construction work. Breathing in asbestos fibres is harmful to the lungs.

INTRODUCTION

Asthma. An illness that causes difficulty in breathing.

Bacteria. Germs that can cause some illnesses.

Barrier cream. A protective cream applied to your hands before starting work.

Bracing. Scaffold poles that make a scaffold rigid.

Brick-guard. A metal mesh fitted to a scaffold to prevent anything from falling through the gaps between the guard-rails and toe-board.

BS. British Standards.

BSI. British Standards Institution.

Cable ramp. A temporary cover laid over a trailing cable to protect it from damage by people or traffic passing over it.

Carcinogen. Substances that can lead to cancer (cancer-causing agents) (for example, asbestos and diesel exhaust fumes).

CCTV. Closed circuit television.

CDM. Construction (Design and Management) Regulations.

Cherry picker. A type of mobile elevating work platform (MEWP) on which a passenger carrying basket is located on the end of an articulating or extending arm.

Chronic. Long-term effects on health (for example, developing asthma from wood dust, or cancer from asbestos exposure).

Consultation. The action or process of having formal talks or discussions (for example, a manager asking you for your input whilst carrying out a risk assessment).

Control measure. Putting measures in place to reduce the risk to an acceptable level (for example, guarding on a machine).

COSHH. Control of substances hazardous to health.

CPCS. Construction Plant Competence Scheme.

Crush injuries. Injuries caused by something crushing a part of the body.

CSCS. Construction Skills Certification Scheme.

Distribution system (electrical). The method that is used to get electrical supplies to where they are needed on site.

Double-handling. Having to move something twice.

EA. Environment Agency.

Edge protection. A framework of scaffold poles or other suitable barrier erected around an open edge to stop anything from falling over the edge.

EMAS. Employment Medical Advisory Service.

Employee. Someone who works for someone else.

INTRODUCTION

Employer. Someone who has people working for him or her.

FRS. Fire and Rescue Service.

HAVS. Hand-arm vibration syndrome.

Hazard. Anything that has the potential to cause harm (ill health, injury or damage).

Health and Safety at Work etc. Act 1974 (HSWA). The main piece of health and safety law.

HFL. Highly flammable liquids.

HSE. Health and Safety Executive.

HSE inspector. An official who can inspect the site and take action if work is not being carried out safely.

Lanyard. A length of fabric that connects a safety harness with a fixed strong-point.

Ligament. A band of tough body tissue that connects bones or cartilage.

LOLER. Lifting Operations and Lifting Equipment Regulations.

LPG. Liquefied petroleum gas.

Method statement. A step-by-step description of how to carry out a job safely.

MEWP. Mobile elevating work platform.

NVQ. National Vocational Qualification.

PAT. Portable appliance testing.

Permit to work. A system used for controlling high-risk activities (such as working on live electrical cables) or where activities need extra controls (such as hot works, confined space entry and breaking ground (permits to dig)).

PPE. Personal protective equipment.

PUWER. Provision and Use of Work Equipment Regulations.

RCD. Residual current device.

RIDDOR. Reporting of Injuries, Diseases and Dangerous Occurrences Regulations.

Risk. The likelihood of an event occurring from a hazard.

Risk assessment. A document identifying the hazards, risks and control measures for a particular activity.

RPE. Respiratory protective equipment.

Scissor lift. A type of MEWP with a platform that rises vertically.

INTRODUCTION

Slewing. Part of an item of plant (such as the jib and counter-weight of a crane) rotating about a vertical axis.

Solvent. Chemical used to dissolve or dilute other substances.

Tripping hazard. Items lying around that you might trip over.

Ventilated. Supplied with fresh air.

WBV. Whole-body vibration.

WEL. Workplace exposure limit.

WWT. Working Well Together.

Further supporting information from CITB

CITB has a wide range of products, publications and courses that could help to improve your health, safety and environment knowledge.

After reading this publication or attending the CITB Site Safety Plus *Health and safety awareness course* (HSA) you may wish to consider the next step in expanding your health, safety and environment knowledge and competence.

Details of course training providers can be found on the CITB website: https://www.citb.co.uk/courses-and-qualifications/find-a-training-course/site-safety-plus-courses/

The Site Safety Plus scheme provides a number of courses that will enhance and develop your skills within the building, civil engineering and allied industries. Courses give everyone, from operative to senior manager, the skills they need to progress.

For further information on Site Safety Plus refer to their scheme rules: www.citb.co.uk/standards-and-delivering-training/site-safety-plus-ssp/scheme-rules/

Feedback

If you have any comments on the content within this product, or suggestions for improvement or extra topics, your feedback would be welcome. You can contact us by email or telephone as outlined below.

publications@citb.co.uk

0344 994 4122

CONTENTS
General responsibilities

What your site and employer should do for you	14
What you should do for your site and employer	14
Introduction	15
What the law requires	15
Providing safe ways of working	16
Sharing information and knowledge	19
How the Health and Safety Executive enforces the law	21

GENERAL RESPONSIBILITIES

01

What your site and employer should do for you

1. Provide a safe place to work.
2. Give you safe plant and equipment, safe substances and materials, safe methods and systems of work, and safe and competent persons to work with.
3. Eliminate or avoid risks where possible.
4. Tell you about the hazards and the risks and how they will be eliminated or controlled.
5. Give you information, instruction and training so you can do your job safely and in safety.
6. Communicate with you and allow you to have your say.
7. Make sure suitable welfare facilities are provided.
8. Make sure personal protective equipment (PPE) and clothing is provided.

What you should do for your site and employer

1. Make sure you fully understand and follow the site rules and your safe system of work.
2. Avoid taking shortcuts or risks.
3. Go to and take part in safety inductions and briefings.
4. Co-operate and get involved.
5. Make sure you fully understand and follow all method statements and precautions.
6. Report anything you think is unsafe.
7. Make sure you wear your protective clothing and use any equipment correctly.
8. Report any damage or faults with protective clothing and equipment.

GENERAL RESPONSIBILITIES

Introduction

Everyone on a construction site is legally and morally responsible for health and safety.

01

By knowing what is needed, you will be able to understand your legal duties, and the legal duties of your employer, for protecting the health and safety of you and other people while at work.

This chapter will help you understand the following.

- What the law requires from everyone.
- Employers must provide a safe place of work.
- Safe systems of work must be planned for, put in place and followed.
- People at work must be trained, competent and involved in protecting their own and other people's health and safety.
- Communication has to be open, effective and honest so everyone knows what is expected of them.
- The Health and Safety Executive (HSE) can and will enforce the law in respect of health and safety.

 Competence

A combination of a person's skills, knowledge, training and experience and their ability to apply these to carry out a task safely. Other factors (such as attitude and physical ability) can also affect competence.

What the law requires

The Health and Safety at Work etc. Act 1974 (HSWA) was introduced to protect the health and safety of everyone at work.

Your employer is responsible for the following.

- Protecting the health, safety and welfare of all their employees at work.
- Providing and maintaining plant and systems of work that are safe.
- Making sure articles and substances are safely used, handled, stored and transported.
- Providing information, instruction, training and supervision in a way that everyone understands.
- Maintaining any place of work under their control in a condition that is safe and without risk to health.

GENERAL RESPONSIBILITIES

The Health and Safety at Work etc. Act imposes a legal duty on you as an employee. This means **you** have **legal duties to do the following.**

- Work safely, by protecting your own health and safety and that of other people who may be endangered by your acts or omissions.
- Co-operate with your employer in relation to health and safety. (Make sure you understand and follow all safety rules, method statements and precautions.)
- Make sure you wear your protective clothing and use any equipment correctly.
- Report anything you think is unsafe (including any damaged or faulty protective clothing or equipment).

 The Health and Safety at Work etc. Act states that employers must give every worker suitable health and safety information, instruction, training and supervision, as is needed to make sure workers are kept safe in the workplace.

Providing safe ways of working

The law requires employers to develop safe systems of work (a defined method of carrying out each job in a safe way).

The following documents can form part of the safe system of work.

- Health and safety policy.
- Risk assessments.
- Method statements.
- Permit to work.
- Construction phase plan.
- Survey results.
- Health surveillance.
- Control of substances hazardous to health (COSHH) assessments.

Bad practice: this image shows unsafe work at height. Make sure you follow safe working methods to prevent accidents

Health and safety policy

Your employer's health and safety policy will give you the following information.

- Explain how health and safety is managed in your company.
- Identify what the arrangements are (what should be done and how it should be done).
- Show who is responsible for doing what (including what you and your fellow workers must do).

GENERAL RESPONSIBILITIES

Risk assessments

Risk assessments explain the following.

- The hazards of the work site or task (for example, an open excavation).
- The significant risks (for example, people falling in).
- The control measures needed to minimise the risk to an acceptable level. (For example, erect a double handrail around all sides of an excavation, with a safe way of getting in and out for workers. Make sure excavations are inspected and that they have suitable barriers, warning signs and are adequately lit at night.)

Your employer should assess the risks to both you **and** others arising from the work being done.

A *hazard* **is anything that could cause harm to you or another person (such as chemicals, working from a ladder, or electricity).**

The *risk* is the chance (likelihood) that you, or someone else, could be harmed by the hazard and how serious any harm could be.

Method statements

Method statements explain how the job is to be done safely. They will also identify the following.

- The sequence, method and controls to be followed.
- The materials and equipment to be used.
- The number of people and the skills, knowledge, training, experience and supervision needed.

Permit to work

A permit-to-work system is used for controlling high-risk activities, or where there are activities that need extra controls. It is a formal, dated and time-limited certificate signed by a properly authorised and competent person. Permits have strict controls and limitations that must be followed. You must never start any job for which a permit is needed before the permit's start time and before the controls are in place.

The types of activity that need a permit-to-work system include the following.

- Working on live electrical cables.
- Hot works or welding.
- Confined space entry.
- Breaking the ground (permit to dig).

GENERAL RESPONSIBILITIES

Construction phase plan

A construction phase plan is needed for every construction project. It must be regularly reviewed and may be added to or changed as the project goes ahead. It will give you the following information.

- The identity and location of the main dangers on site and how they will be controlled.
- How the work has been planned with safety in mind.
- How the site will be organised.
- How people will work safely together.
- The arrangements for providing welfare facilities.

Survey results

There are a number of surveys (for example, an asbestos, noise or underground services survey) that may be needed. These survey results will form part of the safe system of work, as they will show where certain hazards may be.

Health surveillance

Health surveillance allows for the signs of ill health to be identified early. It may be needed if you are exposed to a hazardous agent or if your work may have an adverse effect on your body (for example, hearing damage if you are working in a noisy area). It lets your employer check that their control measures are working and that you are not at risk for any ill health that could be prevented.

COSHH assessments

COSHH assessments are an assessment of health risks created by work involving substances hazardous to health. COSHH assessments concentrate on the hazards and risks from hazardous substances in your workplace. Remember that health hazards are not limited to substances labelled as hazardous. Some harmful substances can be produced by the method of work you use (for example, wood dust from sanding, or silica dust from tile or block cutting).

Employers must carry out a COSHH assessment for the following reasons.

- To identify the hazardous substances used in, or created by, a work process to which you and others will be exposed.
- To establish the degree of risk to your health resulting from exposure.
- To devise safe systems of work that either eliminate exposure or control it to an acceptable level.

GENERAL RESPONSIBILITIES

01

 Do you follow a safe system of work?

Are you familiar with your safe system of work? Do you know what the sequence of work is? Are you aware of the hazards? Do you have the right equipment and training?

Do you fully understand the identified control measures? Or are you doing the job the way you have always done it, the way you think it should be done or just having to make do with what you have got?

If you are working differently from the written and approved safe system of work then speak to your supervisor or employer. Your way may be better, quicker or more efficient but it may have other risks that you don't know or haven't thought about. The system of work will need to be assessed and, if necessary, extra controls put in place.

The law needs you to have the necessary **skills**, **knowledge**, **training** and **experience** so you can safely carry out the tasks given to you. You should never be put in a position where you have to carry out a job for which you do not have the necessary competence, equipment and safe method of working. If you think you are, then you may be putting yourself or others at risk – you need to speak to your supervisor or the site manager.

Sharing information and knowledge

Good communication of health and safety information is essential for everyone.

Employers have a legal duty to consult with their workforce. To be effective it needs to be two way (both sides listening, as well as talking).

This can be done in a variety of ways.

- Site inductions.
- Posters.
- Safety briefings.
- Toolbox talks.
- Informal chats and open door policies.
- Worker involvement schemes.
- Suggestion boxes.
- Health and safety law poster or pocket card.

 Employers must either display a copy of the official health and safety law poster or give every employee a card or leaflet that contains this information.

GENERAL RESPONSIBILITIES

Site induction

You must attend a site induction for each new site that you visit. As a minimum you should be told about the following.

- The site rules.
- The site hazards (for example, overhead power lines).
- Traffic management systems.
- Areas of the site where you can and cannot go.
- Welfare facilities.
- Emergency and first-aid arrangements.
- Personal protective equipment (PPE).
- Permits to work.
- Environmental considerations and any protective requirements.
- How you will be consulted regarding health and safety matters.

Completing an induction record form

Toolbox talks

These are short health and safety briefing sessions that may be on a particular subject connected to the work being carried out, or that give other important safety information and advice.

The person delivering the toolbox talk should give you the opportunity to ask questions or raise concerns.

 The aim of a toolbox talk is to give and share information to keep you safe and protect your health.

Worker involvement

Workers are often the best people to understand the risks in their workplace. Talking to, listening and co-operating with each other can help in the following ways.

- Identify joint solutions to problems.
- Raise standards.
- Reduce accidents and ill health.
- Help people to work safely and in safety.

You may be asked to take part in site inspections or audits and discuss your usual work methods. This will help your employer understand any problems you are experiencing and help them to propose solutions.

GENERAL RESPONSIBILITIES

01

 Have your say

Many sites run suggestion schemes, have regular safety forums or meetings and have open door policies.

Get involved - your views are important and can make a difference.

How the Health and Safety Executive enforces the law

The HSE is a Government body responsible for overseeing most aspects of workplace health and safety in the UK. The HSE's main roles are listed below.

- Offer advice on workplace health and safety.
- Carry out workplace inspections.
- Conduct serious accident investigations.
- Carry out enforcement action for health and safety breaches.

HSE inspectors have the following powers.

- The legal power to demand entry to the workplace without notice, with a police officer if necessary.
- Can interview you or anyone else, with or without caution.
- Can prosecute a company, an employer or an **individual employee**, in a Magistrates' or Crown Court.
- Can raise and issue improvement or prohibition notices, on an employer or an individual worker. Such notices must be complied with.
- Can charge the employer a fee for intervention.
- Can measure, and take photographs, samples or possession of anything, if needed for evidence.
- Can inspect record books and any other documents.
- Can demand that the scene of an accident remains undisturbed.

HSE inspectors can issue notices if they think it necessary

Improvement notices

These are issued if something is unsafe, not up to standard or not being adequately controlled. They will say how, in the HSE inspector's opinion, the law was being broken and give a date by which things must be put right or improved.

GENERAL RESPONSIBILITIES

Prohibition notices

These are issued when, in the inspector's opinion, something is so unsafe that all work connected to it must **stop** immediately (on issue of the notice) and must **not** start again until the matter has been put right.

Fee for intervention

The HSE can charge an hourly rate for the time the HSE inspector spends investigating a breach of health and safety law, including visits, letter writing and ensuring that the matter has been put right.

Examples of HSE prosecutions

 Operator blamed by HSE, as employer had provided proper training

A 25-year-old street lighting operative, employed by a highways maintenance contractor, has been ordered to pay over £5,000 in fines and prosecution costs after failing to put up temporary barriers around the lamp where he was working.

A toddler was injured when part of a streetlight fell as she was passing underneath. The child was in a pram being pushed along a London street when the reflector from a streetlight struck her on the head, causing a wound that needed stitches.

The operative had been told to investigate a faulty street lamp in Hackney, London. When he detached the reflector at the top of the lamppost it fell toward the mother and her daughter below.

Temporary barriers should have been used to separate pedestrians from the work area around the lamp before the operative dismantled the reflector. The operative pleaded guilty to breaching Section 7(a) of the Health and Safety at Work etc. Act 1974. He was fined £2,250 and ordered to pay costs of £2,888.

Speaking after the case, a HSE inspector said: 'This was a serious incident that needlessly injured a small child and caused her mother understandable distress. No blame can be attached to the contractor as the operative had been properly trained by his employer to carry out this kind of work safely. Individual employees must realise that they face criminal prosecution by the HSE if they show a reckless disregard for health and safety, putting others at serious risk.'

GENERAL RESPONSIBILITIES

 Company fined £2.4 million following two corporate manslaughter charges

A construction company director has been jailed after two workers fell from a first-floor balcony at a luxury flat in Knightsbridge, West London. The 44-year-old director of the renovation firm was found guilty of breaching Section 2(1) of the Health and Safety at Work etc. Act and received a 14-month prison sentence for each death, which will run concurrently. He was also barred from being a company director for four years.

The company, which has gone into liquidation since the incident, despite previously having an annual turnover of around £9.7 million, was found guilty of two counts of corporate manslaughter and two breaches of Section 2(1) of the Health and Safety at Work etc. Act. The company was fined £1.2 million for each death and £650,000 for Health and Safety at Work etc. Act breaches, all of which apply concurrently. It must also pay £72,000 costs.

The court heard that in November 2014 two Polish nationals, aged 22 and 29 years, fell from the balcony of a flat in London's Cadogan Square, which was being refurbished by the renovation company. The men were part of a group of five workers who were using ropes to haul a 115 kg sofa up 6 m, over a balustrade and onto a balcony, with only the iron Victorian railings of the balcony acting as a barrier. The 130-year-old railings gave way and the two men fell from the balcony to the ground. The 22-year-old was pronounced dead at the scene, while the 29-year-old was taken in a critical condition to a central London hospital, where he later died of his injuries.

The sofa delivery company had recommended that an external furniture lift should be hired for the lift and had emailed the director of the renovation company with an estimated budget of £848 to hire an external lift. The director responded with a message saying: 'Unfortunately, we do not have time for all that. Please deliver the sofa and we will get it up to the flat'.

The work on the flat was over budget and behind schedule. The budget had increased from around £650,000 to as much as £920,000. The court heard that on the day of the accident there was a failure to identify who was supervising the site. The prosecution told the court that none of the training documents provided were in Polish, despite the workers not speaking English, suggesting that the documents were there 'just for show'. The men were also not provided with a plan, method statement or risk assessment before the task started.

The judge stated that the director's motive for ignoring the warnings about the furniture lift: 'must have been in one way or another to benefit his business. The word has got to get out that health and safety on building sites is not a boring technicality. It is vital to the safety of employees and others in what is inherently a dangerous environment. Those who are willfully blind to the risks, despite warnings – as you were – have got to expect to go immediately to prison.'

(Source: HSE.)

GENERAL RESPONSIBILITIES

 Builder sentenced after young worker is seriously injured

Magistrates ordered a builder from Cornwall to pay out nearly £10,000 after an employee sustained serious, life-changing hand injuries whilst operating a hand-held circular saw.

Bodmin Magistrates' Court heard how the building services company was refurbishing a barn in Callington in February 2017. A young worker, who had just turned 17, was using a circular saw to cut wooden flooring sheets when the blade made contact with his hand. It cut right through his index finger, three-quarters of the way through his middle finger and halfway through his ring finger.

An investigation by the Health and Safety Executive (HSE) found that the builder had no record of any information, instruction and training that he had provided to his employee in the safe use of the circular saw, and nor had he ensured that safe working practices were followed when cutting the flooring sheet.

The investigation also found that the circular saw blade had not been properly adjusted for the size of material being cut at the time of the incident and the flooring sheet was not appropriately supported while being cut.

The accused builder pleaded guilty to breaching Section 2(1) of the Health and Safety at Work etc. Act 1974 and Regulation 3(4) of the Management of Health and Safety at Work Regulations 1999. He was fined £1,120 and ordered to pay costs of £8,489.48.

An HSE inspector said after the hearing: 'This injury was easily preventable and the risk associated with the task should have been identified. Employers should make sure they properly assess and apply effective control measures to minimise the risk from contact with dangerous parts of machinery to ensure that the risks are given careful attention to ensure they are properly controlled.'

CONTENTS
Accident reporting and recording

What your site and employer should do for you	26
What you should do for your site and employer	26
Introduction	27
Common types of accident and incident	27
Prevention – what you can do	27
Reporting and recording	28
Occupational diseases	29

ACCIDENT REPORTING AND RECORDING

What your site and employer should do for you

1. Put measures in place that allow different trades to work safely together.
2. Make sure all employees fully understand the procedures for reporting accidents, incidents and near misses.
3. Provide an accident book and make sure records are kept of all accidents.
4. Make sure that certain types of serious injury, incidents and occurrences of certain types of work-related ill health are reported to the Health and Safety Executive (HSE), where required.
5. Listen, and make sure any concerns you have are acted on.
6. Make sure accident and incident investigations are carried out to understand what went wrong and stop it from happening again – to find the cause, not to find someone to blame.

What you should do for your site and employer

1. Understand and follow your safe system of work.
2. Stop and take advice if you cannot follow your safe system of work.
3. Report any accident you have to your manager or supervisor and make sure the details are recorded in the accident book.
4. Report any near misses or anything you think could be unsafe.
5. Co-operate with any investigation.

ACCIDENT REPORTING AND RECORDING

Introduction

 On average 35 construction workers are killed each year due to accidents.

Over half of the fatal injuries to workers are from three types of incident.

1. Falls from height.
2. Being struck by a vehicle.
3. Being struck by a moving or falling object.

Year after year the same types of accidents and incidents are repeated.

Common types of accident and incident

- Falling from height.
- Slips, trips and falls on the same level.
- Manual handling.
- Being struck by mobile plant.
- Contact with electricity.
- Contact with moving machinery.
- Being trapped by something collapsing.
- Being injured in overturning vehicles and plant.

In many cases these accidents could have been avoided by taking simple precautionary measures (for example, clearing up the work area regularly to reduce the likelihood of slips, trips and falls).

Prevention – what you can do

- Make sure you fully understand the safe system of work.
- Work to the instructions you are given.
- Follow the site rules.
- Do not be tempted to take short cuts.
- Keep your work and storage areas tidy (good housekeeping).
- Keep access routes and walkways clear of materials and equipment.
- Report anything you think may be unsafe to your supervisor.
- Stop work if you think it is not safe.

ACCIDENT REPORTING AND RECORDING

All accidents and incidents should be reported. Your employer can then investigate, find out the cause and prevent them from happening again.

 Everyone on site has a responsibility to report unsafe conditions.

Reporting and recording

You must make sure that any accident or injury you have is reported and recorded in the accident book. Unsafe conditions and near misses should also be reported, so action can be taken to prevent them from becoming an accident. Certain types of serious injury and injuries where an employee, or self-employed person, is off work or unable to perform their normal work duties for more than seven consecutive days, have to be reported by your employer to the Health and Safety Executive (HSE).

Unsafe conditions

Something with the potential to cause harm

Near misses

An incident that nearly resulted in an injury or damage

Accidents

An accident that caused an injury or damage

The following details **must** be recorded in the **accident book**.

- The injured person's name and address.
- The injured person's occupation.
- Date and time of the accident.
- Where the accident happened.
- How the accident happened.
- The injury that was sustained.
- Details of the person filling in the book (if different from the injured person).

Your employer must keep accident book information secure, as it is confidential. Most accident books have perforated pages, which can be detached and stored in privacy and allow the injured person to have a copy of their individual information.

ACCIDENT REPORTING AND RECORDING

If you see something that is unsafe or witness a near miss it is important you tell your supervisor or employer. You may think reporting it will get you into trouble – but the opposite is true. If more unsafe conditions and near misses are reported, it is likely there will be fewer accidents.

Occupational diseases

You must tell your employer if you develop signs or symptoms of the following conditions, or if a medical professional tells you that you are suffering from a condition or disease that they think may be related to your work. Certain conditions and diseases have to be reported by your employer to the HSE, such as those listed below.

- Hand-arm vibration syndrome (HAVS) (for example, from using hand-held vibrating tools).
- Severe cramp in the hand or forearm (where the work involves long periods of repetitive movement of the fingers, hand or arm).
- Carpal tunnel syndrome (for example, from using vibrating or percussive tools).
- Tendonitis or tenosynovitis in the hand or forearm (for example, from repetitive or frequent movement).
- Occupational dermatitis (for example, from working with cement).
- Occupational asthma (for example, from breathing in construction dust, such as softwood dust).
- Occupational cancer (for example, from exposure to asbestos fibres or silica dust).
- Disease caused by exposure to a biological agent (for example, leptospirosis (Weil's disease)).

ACCIDENT REPORTING AND RECORDING

02

CONTENTS
Health and welfare

What your site and employer should do for you	32
What you should do for your site and employer	32
Introduction	33
Work-related ill health issues	33
Diseases carried in the blood	35
Stress and mental health at work	36
Drugs and alcohol	37
Welfare facilities	40

HEALTH AND WELFARE

What your site and employer should do for you

1. Identify work activities that could damage your health.
2. Tell you about the hazards and risks and how they will be controlled.
3. Provide suitable equipment and safe methods of working to minimise any exposure.
4. Give you personal protective equipment (PPE), free of charge, where needed.
5. Provide and maintain good, clean welfare facilities.

What you should do for your site and employer

1. Understand how the hazards can damage your health.
2. Use the equipment provided properly and follow the safe system of work.
3. Wear your PPE properly, and report any defects in your PPE.
4. Do not misuse the welfare facilities. Help to keep them clean and tidy.
5. Report any changes in your health or if you think that your work is making you unwell.
6. If you have any concerns or any doubts about anything related to your work, raise the matter with your supervisor or manager.

HEALTH AND WELFARE

Introduction

Work-related ill health can have a serious impact on individuals and their families, which is often misunderstood and underestimated.

This may be because the effects of exposure to most work-related ill health hazards are not always immediate (unlike an accident which causes an immediate injury). You may go home from work feeling more or less the same each day and be unaware of the effects.

If you are repeatedly exposed to small doses of dust, fumes or certain substances, exposed to loud noises or use vibrating tools, they may cause damage to your body. It can take many weeks, months or even years before symptoms of exposure and damage become a problem to you and by then irreversible damage may have occurred.

Work-related ill health issues

Respiratory (breathing) diseases

 Refer to Chapter B07 for information on respiratory diseases.

Noise and vibration

 Refer to Chapter B08 for information on noise and vibration.

Skin conditions

Occupational dermatitis

This is a skin condition that is generally caused by exposure to irritants contained in chemicals and other harmful substances. Bricklayers, stonemasons, and painters and decorators are the trades that are most at risk of occupational dermatitis in the construction industry. The hands and forearms are usually affected, but in some cases the face, legs, feet and other parts of the body can be affected. The condition cannot be passed on from the sufferer to other people. It can be painful and will affect your personal life as well as your work life.

Some of the symptoms are shown below.

- Redness.
- Itching.
- Dryness.
- Cracking or blistering.

The most common cause of dermatitis in the construction industry is working with cement. Wet cement can also cause burns.

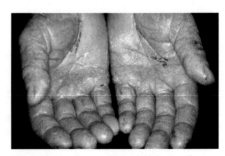

Dermatitis showing crusting and thickening of the skin

HEALTH AND WELFARE

There are two general types of occupational dermatitis, as outlined below.

1. **Irritant contact dermatitis.** This is usually caused by the skin coming into contact with an irritant substance, which is often a chemical or dust. Anyone can be affected. The amount of time a person is exposed, together with the strength of the irritant substance, will affect the seriousness of the complaint. Most cases of dermatitis are of this type.

2. **Allergic contact dermatitis (sensitising dermatitis).** Some people develop an allergic reaction to a specific substance. This reaction may follow after weeks, months or even years of use or exposure to a substance without any ill effects. However, once sensitising dermatitis has occurred, any future exposure to, or contact with, the substance will again produce an allergic reaction. The skin's reaction to irritants varies from one individual to another. The reaction may be only a mild redness, or it can develop into swelling, blisters and septic ulcers that are both unsightly and painful.

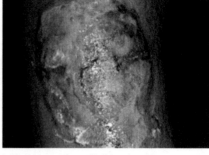

Irritant contact dermatitis 'pizza knee' from a cement burn

Personal hygiene is particularly important when working with materials that may be irritants. It is equally important that clothing is kept clean (for example, oil-stained and dirty overalls are a well known cause of skin problems around the groin).

 Allergic dermatitis can make your skin so sensitive you will not be able to use some substances again, which could end your career.

Skin cancer

Ultraviolet (UV) radiation from the sun is a major cause of skin cancer. Around 800 construction workers are diagnosed with skin cancer every year as a result of sun exposure. People whose jobs keep them outdoors for long periods of time (such as construction workers) may, if their skin is unprotected, get more sun on their skin than is healthy for them. They will then be at a greater risk of developing skin cancer.

The following actions will help to prevent overexposure and reduce the likelihood of skin cancer.

- Covering up exposed skin by wearing suitable loose, long-sleeved and long-legged clothing and a hat.
- Using sun block on exposed arms, face and neck.

 If you notice new warts or moles, changes to existing moles or they begin to itch, seek medical advice as soon as possible. Don't ignore it – early diagnosis and treatment are vital.

HEALTH AND WELFARE

Although less common, skin cancer can be caused by contact with mineral oils. Daily contact with items such as oily clothing or gloves can lead to a form of skin cancer. Mineral oils are common on mechanical plant and pipe-threading machines.

Diseases carried in the blood

Leptospirosis (Weil's disease)

- This disease is caused by bacteria that are present mainly in the urine of infected rats. It can enter the bloodstream through cuts and grazes, or from hand to mouth contact (for example, if the bacteria are present on your hands when eating or smoking).

- It is a particular problem when working on or near water, sewage, waterlogged sites or rat infested areas.

You can minimise the risks by taking the following precautions.

- Wear suitable gloves.
- Keep all cuts and grazes covered with a waterproof plaster.
- Wash your hands before eating or smoking.

Discourage rats from coming onto site; use the welfare or canteen facilities provided and put all food waste in covered bins.

Another form of leptospirosis can be caught from the urine of infected dairy cattle. Anyone involved in construction work on dairy farms must be made aware of this hazard.

 The early symptoms of leptospirosis (Weil's disease) are like influenza (flu). If left untreated the disease may lead to kidney problems and can be fatal, sometimes causing death within three to six weeks.

Tetanus (lockjaw)

- This is a disease of the nervous system.
- It enters the body through cuts, grazes or puncture wounds.
- The bacteria that cause tetanus are found in contaminated soils or manure.
- An early symptom is an increasing difficulty in opening the mouth or jaw.

Hepatitis

Hepatitis is usually caught from contact with infected needles and syringes. A needlestick injury is an accidental puncture of the skin by a hypodermic needle. If not handled in a safe manner, discarded needles can pose serious health risks. Blood on a needle could be infected with a blood borne virus (for example, hepatitis B or C or HIV).

Treat discarded needles with care

HEALTH AND WELFARE

If you find any suspected drug taking equipment leave it alone and contact your supervisor or employer.

Legionella

This is a form of pneumonia, which can be caught by breathing in contaminated water vapour or mist (for example, from air conditioning and cooling towers or hot and cold water systems).

A safe system of work is needed if a water system is suspected of being contaminated by legionella bacteria.

 If you think you may be suffering from tetanus, hepatitis or legionella, get medical advice urgently and tell your supervisor.

Stress and mental health at work

Stress is the adverse reaction people have to too much pressure or other demands placed upon them. It is not an illness in itself, but can lead to you not being able to perform at work and can have an impact on your mental and physical health and wellbeing. Stress can lead to anxiety and depression, as well as to other chronic health conditions such as strokes, back pain, headaches or migraine, stomach problems and drug and alcohol dependency.

Work-related stress and mental health conditions are closely linked, with similar signs and symptoms. Work-related stress can trigger or worsen an existing mental health condition. Your employer should identify any signs of a mental health problem quickly and efficiently so they can help you to get the support or treatment you need.

The early signs and symptoms of a mental health issue may include, but not be limited to, the following.

- Erratic or changed behaviour.
- Poor quality of work.
- Poor time keeping.
- Withdrawal from social contact.
- Increased absenteeism.
- Unusual shows of emotion.
- Changed appearance.
- Increased use of alcohol, drugs or smoking.
- Over performance – people pushing themselves to excess.

Stress can affect you physically and mentally

The earlier it is identified that an individual may be experiencing a mental health problem, the better it will be for all involved. Whatever the cause, if anyone is having problems, it's important that they get help as soon as possible.

HEALTH AND WELFARE

The following may help to improve your mental health.

- Talk to a family member, a friend you trust, a mate at work, or someone independent. Talking about how you feel is not a sign of weakness and having someone listen to you can often make you feel better.

- Look after your physical health. Good mental and physical health are directly linked. Physical activity, sleep and diet can all impact on your mental health. Getting a good night's sleep and eating a healthy diet can all improve the way you feel.

Talking about how you feel is not a sign of weakness

- Do things that you enjoy (for example, a hobby, reading or sport) and try to avoid things that you don't.

- Recognise your limits, as this will help you recognise the things that are likely to have a negative impact on your mental health. Stop when you need to, and take a break if you feel like things are getting too much.

 Many sources of help are available. For further information refer to organisations such as Mind (www.mind.org.uk), Time to Change (www.time-to-change.org.uk), Bipolar UK (www.bipolaruk.org) and Heads Together (www.headstogether.org.uk).

Drugs and alcohol

Some medicines you can buy over the counter (such as hay fever medicines or cold remedies) as well as some prescription medicines can make you drowsy or have other side effects. This can affect your overall ability and judgement and mean you are not safe to be on a construction site and not fit to drive or operate machinery. You should read the label or speak to the pharmacist if in doubt. You must let your employer know if you have to take any medicines at work.

Many employers and clients have policies to carry out random testing for drugs and alcohol.

Anyone caught working under the influence of illegal drugs, some types of medicine or medication, psychoactive substances (known as 'legal highs') or alcohol may have to leave site immediately and could lose their job.

HEALTH AND WELFARE

Effects of drugs and alcohol

Regardless of if it is a medicine, alcohol or any type of drug, any person under the influence of any substance, while at work, is a danger to themselves and everyone else on site. They are likely to suffer from the following.

- Poor or irrational decision making.
- Slow reaction times.
- Clumsiness.
- Distorted vision.
- Mental confusion.

The effects and traces of some drugs can stay in the body for long periods, even after users think the effects have worn off (for example, cannabis can be detected during testing several weeks or even months after it has been taken). Unlike alcohol, there are no safe levels for illegal drugs so a trace could mean failing a drugs test. This includes drugs referred to as 'legal highs' (psychoactive substances).

Psychoactive substances

Psychoactive substances are designed to produce similar effects to drugs (such as cocaine and ecstasy). They were known as 'legal highs' until changes to the law in 2016, making them illegal under the Psychoactive Substances Act.

In 2016 there were 123 deaths involving psychoactive substances. Around 80% of deaths were males. Most deaths involving psychoactive substances happen in people aged 20 to 29 and the average age for deaths of this type is 28 years old.

 It is an offence under the Psychoactive Substances Act to possess with intent to consume, produce, offer to supply, supply, import or export any substance for human consumption that is capable of producing a psychoactive effect. Offences are punishable by a fine or up to seven years' imprisonment.

Illegal drugs

The maximum penalties for illegal drug possession, supply (dealing) and production depend on the type or class of drug. Examples of maximum penalties are outlined in the table below.

Class	Examples of drugs	Possession	Supply and production
A	Crack cocaine, cocaine, ecstasy (MDMA), heroin, LSD, magic mushrooms, methadone, methamphetamine (crystal meth)	Up to seven years in prison, an unlimited fine or both	Up to life in prison, an unlimited fine or both

HEALTH AND WELFARE

Class	Examples of drugs	Possession	Supply and production
B	Amphetamines, barbiturates, cannabis, codeine, ketamine, methylphenidate (Ritalin), synthetic cannabinoids	Up to five years in prison, an unlimited fine or both	Up to 14 years in prison, an unlimited fine or both
C	Anabolic steroids, benzodiazepines (diazepam), gamma hydroxybutyrate (GHB), gamma-butyrolactone (GBL), piperazines (BZP)	Up to two years in prison, an unlimited fine or both (except anabolic steroids – it's not an offence to possess them for personal use)	Up to 14 years in prison, an unlimited fine or both

Alcohol

Depending upon your employer or site policy, the blood-alcohol level needed to pass a test can be less than the legal driving limit.

Besides being a danger to themselves and others, anyone under the influence of alcohol at work risks losing their job and livelihood. Many people convicted of drink driving were driving the morning after drinking the night before. Most people seriously underestimate the length of time it takes to sober up.

 The effects of alcohol can last for hours

On average it takes one hour for your body to get rid of one unit of alcohol.

A unit is the volume (litres) x ABV (% alcoholic strength of the drink).

- One pint of beer at 5% strength is 2.8 units.
- Five pints of beer at 5% strength is 14 units.
- One 200 ml glass of wine at 13% strength is 2.6 units.
- One 25 ml shot of a spirit at 40% strength is 1 unit.
- One 275 ml bottle of alcopop at 5% strength is 1.4 units.

If someone started drinking at 19:00 hrs and drank five pints of beer with a strength of 5% it could be up to 10:00 hrs the next morning before all the alcohol is out of their system. However, this time can change depending on things like your size, weight, age, when you last ate and how much sleep you have had.

 The Department of Health recommends that alcohol intake is limited to 14 units per week (two units per day) for men and women.

HEALTH AND WELFARE

 There is no safe limit of alcohol that you can consume and then drive or operate machinery safely.

 For further information on alcohol visit the Drinkaware website (www.drinkaware.co.uk/alcohol-facts/).

Welfare facilities

Your employer has a legal duty to provide adequate welfare facilities, which include the following.

- An adequate number of separate toilets for men and women. Where this isn't possible, toilet doors should be lockable and separate from urinals.
- Toilet paper.
- Sanitary disposal facilities for women.
- Adequate and suitable hand-washing facilities (not the canteen or kitchen sink), to include running hot (or warm) and cold water, soap or hand cleaner, and a way of hygienically drying the hands (not a shared towel).
- Showers, if necessary. (Where showers are necessary there should be separate showers for men and women, unless it is a single shower with a lockable door.)
- Changing and drying rooms, with separate facilities for men and women, where necessary.
- Somewhere to take breaks from work, with a supply of clean drinking water and cups, or a drinking water fountain.
- A means of boiling water for drinks.
- Tables and chairs with back support (not canteen benches).
- A facility to warm up food (for example, a microwave).
- A means to secure valuables and a change of clothes (for example, lockers).

Welfare facilities must be kept clean and in good order. If they are not, then speak to your supervisor or the person responsible.

 The same level of welfare facilities must be made available for transient (mobile) workers.

 Welfare facilities are provided for your benefit – please look after them.

Site canteen area for taking breaks

CONTENTS

First aid and emergency procedures

What your site and employer should do for you	42
What you should do for your site and employer	42
Introduction	43
First aid	43
Discovering a casualty	43
What employers must provide	44

FIRST AID AND EMERGENCY PROCEDURES

What your site and employer should do for you

1. Tell you the first aid and emergency procedures during the site induction.
2. Maintain emergency escape routes and equipment.
3. Provide first aiders and develop first-aid procedures.
4. Display emergency information, contacts and telephone numbers.
5. Provide emergency and rescue equipment.
6. Make accident books available and make sure records are confidential once completed.

What you should do for your site and employer

1. Be aware of the emergency procedures relating to your own workplace and safe system of work.
2. Know how to raise the alarm.
3. Know where and how to get first aid.
4. Know what to do in an emergency or site evacuation.
5. Know where to gather if there is an emergency.
6. Make sure the details of any accident you have are entered into the accident book.

FIRST AID AND EMERGENCY PROCEDURES

Introduction

Every site should have procedures in place in case of an emergency such as fire, serious injury, collapse, people trapped or needing rescue, or chemical spill.

You must know the following information.

- How to raise the alarm.
- What the alarm sounds or looks like.
- Where to go and what to do.
- The escape or evacuation route.

Information should be given during site induction and displayed on noticeboards or signs.

Remember – sites are constantly changing and so can emergency escape routes.

 Make sure you always know your escape route and where any equipment you may need is kept.

You should also be familiar with any emergency procedures that form part of your safe system of work (such as confined space rescue and rescue from height).

First aid

During your site induction the following information should be explained.

- Where to find the first-aid kit.
- Who the first aider and/or appointed first-aid person is.
- How to contact or recognise the first aider and/or appointed person (for example, green hat or sticker on hard hat).
- Where to get first-aid treatment.

First-aid post

The primary aims of first aid are to preserve life (by carrying out emergency first-aid procedures), prevent the casualty's condition from worsening (for example, by making the area safe or keeping the casualty from moving) and promote recovery (by calling an ambulance and arranging fast emergency medical help).

Discovering a casualty

If you are first on the scene of an accident your actions could be crucial.

You should do the following.

- Assess the situation.
- **Make sure you do not put yourself in danger.**
- If it is safe to do so, remove or isolate the hazard.
- Go to the casualty and find out what is wrong.
- **Call for help** – if no-one comes explain to the casualty that you are going to go and find help and call the emergency services.
- Return and stay with the casualty until help arrives.

 It is crucial that time is not wasted. The priority is getting help (first aid) and the emergency services to the injured person as soon as possible.

FIRST AID AND EMERGENCY PROCEDURES

What employers must provide

Accidents and injuries do happen on site and employers must provide the following.

- Adequate first-aid cover at all times when the site is in operation, including shift work, which may mean appointing trained and competent people to act as first aiders or emergency first aiders, and making sure they are available at all times the site is occupied.
- Sufficient and suitable first-aid equipment for the hazards and the risks identified (for example, eyewash and burns kits).
- Arrangements for lone workers or, when the first aider is not there, a means to call for an ambulance or other professional help.

Employers should also take into account how far the site is from a hospital or the emergency services.

First-aid roles

Based on the findings of their assessment of first-aid needs, an employer could provide first-aid cover by appointing one or more of the following roles.

Appointed person first aid – is appointed when a first aider is not needed in the workplace. Appointed persons are not allowed to give first aid. Their job is to look after the first-aid equipment and facilities and call the emergency services if needed.

Emergency first aider – trained to give emergency first aid at work (EFAW) to someone who is injured or becomes ill.

First aider – trained in first aid at work (FAW) and able to provide first aid for a greater range of injuries and illnesses to someone who is injured or becomes unwell at work. Sometimes there may be other hazards on site (for example, working in confined spaces), therefore extra training may be needed.

 If you use any first-aid equipment you must inform the person responsible for managing it: the first aider, the appointed person or your supervisor.

Mental ill health and first aid

Following the first-aid needs assessment, your employer might decide that it will be beneficial to train personnel so that they can identify and understand symptoms and are able to support someone who may be experiencing a mental health issue.

First-aid training courses covering mental health teach delegates how to recognise warning signs of mental ill health and help them to develop the skills and confidence to approach and support someone, while keeping themselves safe.

A mental health first aider is someone who has undertaken training and attained a recognised qualification as a mental health first aider (MHFA).

CONTENTS
Personal protective equipment

What your site and employer should do for you	46
What you should do for your site and employer	46
Introduction	47
Policies and site rules	47
Types of personal protective equipment	48
Types of respiratory protective equipment	53

PERSONAL PROTECTIVE EQUIPMENT

What your site and employer should do for you

1. Identify hazards and risks and completely remove them where possible.
2. Where there is still a risk, provide personal protective equipment (PPE), which protects you from those risks.
3. Show you how to correctly wear and care for your PPE, and where it should be worn.
4. Supply PPE free of charge.
5. Provide facilities to clean, store and maintain PPE.

What you should do for your site and employer

1. Wear your PPE at all times when needed.
2. Wear the right PPE for the task, wear it properly and make sure it is properly adjusted.
3. Look after your PPE as instructed. (Care for it and it will care for you.)
4. Report any PPE defects. Damaged, worn out or ill-fitting PPE will not protect you.

PERSONAL PROTECTIVE EQUIPMENT

Introduction

Personal protective equipment (PPE) consists of items of equipment and clothing designed to protect you from a variety of hazards. The type of PPE that is designed to protect you from respiratory (breathing) hazards (such as the inhalation of dust and fumes) is called respiratory protective equipment (RPE).

Your employer must make sure of the following.

- A risk assessment is carried out to identify any hazards and risks, and what can be done to remove or control them without using PPE.
- You are supplied with PPE free of charge, told why it is needed (whenever there is a hazard that cannot be removed or minimised to a safe level by other methods), how to adjust, fit and wear it, and what to do if it is damaged or lost.
- You use the PPE supplied as and when necessary and in line with the information and instruction given.

 PPE should only be used as a last resort.

Policies and site rules

Some sites have PPE policies or rules where you must use and wear head, eye and hand protection. This is because site workers and visitors can be exposed to a variety of hazards and risks, which can change daily and can be outside the control of the employer or site management.

On most sites you can only take off PPE when you are in a safe area (such as the site office or welfare compound).

Some common examples of PPE are listed below.

- Safety helmets (hard hats).
- Safety footwear.
- High-visibility clothing.
- Safety glasses (low-impact eye protection).
- Gloves.

Wearing a combination of PPE

There are times when you will need to wear other extra PPE, as identified in the following circumstances.

- From site rules or site induction.
- Instructions from your supervisor or employer.

PERSONAL PROTECTIVE EQUIPMENT

- As a result of a risk assessment or a direction in a method statement.
- As directed by signs and notices.

Other task-specific PPE is listed below.

- Impact-resistant eye protection (for example, safety goggles when using a disc cutter).
- Ear defenders (muffs) or earplugs.
- Safety harnesses and lanyards.
- Knee pads and overalls.
- Wet and cold weather clothing.
- Lifejackets or buoyancy aids.

Task-specific RPE includes half or full mask respirators.

Your employer must provide the PPE, but you must be responsible for it and for wearing it properly.

- Take care of your PPE (keep it clean and inspect it regularly).
- Use your PPE as instructed.
- Stop work and report any lost or damaged PPE to your supervisor.
- **Do not** work without PPE where it is needed.

Types of personal protective equipment

Head protection

Safety helmets must be worn at all times, except when you are in an area declared as a safe area by site management.

A safety helmet is worn to protect you from falling objects or bumping your head. Your safety helmet will be most effective when you follow the guidelines below.

- Wear it the right way round (peak at the front).
- Adjust it so it fits snug and square on your head.
- Make sure it is fitted and worn with a chin strap if there is a risk of it falling off while you are working.
- Make sure it is fitted with a proprietary liner for cold weather, as required, (do not wear it over your woolly hat).

Safety helmet to help protect against head injury

PERSONAL PROTECTIVE EQUIPMENT

Do not cut, drill holes, paint or apply unauthorised stickers to your safety helmet, as this can severely reduce its capacity to protect you in an incident.

 Do not wear hoodies or beanies under your safety helmet. If extra comfort is needed to keep you warm an approved manufacturer's lining should be used.

Dropping your safety helmet from height onto a hard surface will reduce the strength, even if there is no obvious damage. If this happens it should be replaced.

Foot protection

Safety footwear must be worn at all times on site and should have protective toe-caps and a steel mid-sole or other form of protection to protect from puncture injuries (such as standing on a nail).

- Some offer better ankle support.
- Some (such as safety trainers) offer good grip on slippery surfaces.
- Some offer increased comfort and are more suitable to trades (such as floor layers) who repeatedly kneel and bend their feet.

High-visibility clothing

- All staff and visitors on site should wear a high-visibility vest or coat as a minimum. The colour of the background material should normally be flourescent yellow and the reflective material should comply with EN ISO 20471.

- There are three classes of high-visibility clothing.

    Class 1 – low visibility (tabards and waistcoats) suitable for general sites.

Always wear appropriate hi-vis clothing

Class 2 – medium visibility needed when working on or near A and B class roads or sites with vehicle or plant movement. (For example, vest with two 5 cm bands of reflective tape around the body or one 5 cm band around the body and braces over both shoulders.)

Class 3 – high visibility needed when working on or near dual carriageways, motorways or airports. (For example, long-sleeved jacket and trouser suit with two 5 cm bands of reflective tape around the body and arms, and braces over both shoulders.)

PERSONAL PROTECTIVE EQUIPMENT

Body protection
Protective clothing can be used to protect against the following.

- Strong oils and chemicals (for example, cement).
- Rough or sharp surfaces.
- Ignition and burning (fire retardant).
- Extreme cold (thermal clothing) or heat (long-sleeved, lightweight clothing).
- Extreme (wet and windy) weather.

Hearing protection
There are two main types of hearing protection.

1. Ear defenders or earmuffs.
   - These are worn around the shell of the ear and have to be a snug fit to be effective.
2. Earplugs.
   - These are inserted into the outer ear.
   - Make sure your earplugs are inserted properly. They should not feel loose or fall out.
   - Do not reuse disposable earplugs, as this can cause infection.

Eye and face protection
The three main types are shown below.

Safety glasses (low-impact eye protection)   Impact-resistant safety goggles   Face shield

Safety glasses and other forms of low-impact eye protection are not designed for tasks where high-impact eye protection should be worn (such as when using cartridge tools, grinders or disc cutters).

PERSONAL PROTECTIVE EQUIPMENT

- Different types of eye protection protect you against the following.
  - Flying debris and objects.
  - Chemical splashes.
  - Airborne dust.
  - Molten metal and sparks.
- Eye protection needs to be regularly cleaned and stored to protect it from scratching.
- You need to be able to see through your eye protection for you to be able to work safely.
- If your eye protection is scratched or mists up then you need a suitable replacement.

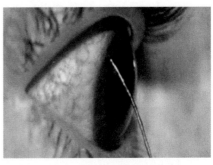

Eye injury sustained as a result of not wearing eye protection

 Safety glasses are only classed as low-impact eye protection and will not withstand an impact from flying debris when using grinders, disc cutters, cartridge tools, and so on. *You are only ever going to have one pair of eyes – make sure you look after them.*

Hand and skin protection

- Gloves should be suitable for the hazards and task. Using the correct type of gloves will protect your hands.
- If using chemicals the gloves should be impervious (the chemical should not go through the glove onto your hands).
- It is important that gloves are regularly cleaned or replaced.

 5%–10% of construction workers working with cement, mortar and concrete are affected by dermatitis.

Barrier creams are no substitute for gloves but can offer some extra protection. Using hand soap, hand cleaners and after work creams to replace oils lost from your skin will help in prevention.

Never use solvents or spirits to clean your hands. These strip the protective oils from your hands leaving you more prone to skin damage.

Protection from sun exposure

You should be careful whilst working outdoors in summer, particularly in the three or four hours around midday when the sun is most intense. You should protect yourself from too much sunlight; using protective clothing to cover up is one of the main ways to avoid the danger. Some other ways are listed on the following page.

PERSONAL PROTECTIVE EQUIPMENT

- Wear close-woven fabrics (such as jeans) and long-sleeved shirts, which will help to protect you from ultraviolet (UV) rays.
- Try to work and take breaks in the shade.
- Wear a hanging flap on the back of your safety helmet to protect your neck.
- Keep a shirt or other top on.
- Use sun block on exposed areas of your arms, face and neck.
- Check your skin for signs of change or damage.
- Drink plenty of water to prevent dehydration.

Site sun block cream dispenser with details of the day's UV level

Fall protection

Items of equipment that are used by a person to prevent that person falling from height (as in the examples below) are also classified as PPE.

- Safety harness.
- Fall-arrest or restraint lanyard.
- Inertia reel fall-arrest block.

All employer and employee duties that apply to other types of PPE also apply to this type of equipment.

A major consideration for planning and the use of this type of PPE is the quick rescue of anyone who has fallen and is suspended in a harness. If a worker has fallen and their fall has been arrested, any delay in rescuing them could result in suspension syncope or fainting. There are many kits available to rescue those suspended at height.

Worker using fall restraint to undertake maintenance work

PERSONAL PROTECTIVE EQUIPMENT

Personal flotation equipment, including lifejackets

- If you have to work on, over or near to water there is a risk of drowning and you should be provided with personal flotation equipment (such as a lifejacket). This should be worn properly at all times that you are at risk of falling into the water.
- Falling into water wearing your work clothing and boots, or even your tool belt, will drag you under or sap your energy quickly, especially in cold or fast moving water.
- Self-inflating lifejackets automatically inflate and turn you onto your back, even if you are unconscious, allowing your airway to open, helping you to breathe.
- Other, more basic but equally important equipment (such as lifebuoys, life-rings or throwing lines) may also be provided.
- Where strong currents or fast flowing water are present there may be a need for other equipment (such as manned rescue boats).

Types of respiratory protective equipment

To prevent exposure to harmful dust, fibres, fumes and gases, employers must provide you with the correct type of respiratory (breathing) protective equipment (RPE).

The choice of RPE will depend upon the nature of the hazard from which you need to be protected. In many cases RPE will only protect against one type of hazard (for example, dust), although combined protection RPE is also available. Colour coding is used to show what the equipment's filter will protect against. The choice of RPE must be made by an authorised, competent person. The wrong equipment or filter will offer no protection.

As with all PPE you should be told why it is necessary, how and when to wear it, how to keep it clean and store it, and what to do if you have any doubts about its effectiveness or if it gets lost or damaged.

If you are issued with RPE, your employer should arrange for you to be face-fit tested to make sure the RPE fits your face properly and functions correctly.

- Where dust cannot be avoided you must wear suitable RPE (masks or respirators).
- Any mask you wear should have a British Standard (BS EN) number and/or CE mark printed on it.
- Disposable masks should have at least two adjustable straps and may have a flexible nose band to shape around the bridge of your nose.
- There should be a good seal between the facepiece and your skin. Facial hair and other items of PPE (such as safety glasses) can interfere with the fit.

Cheap, disposable (nuisance) dust masks from DIY stores do not meet PPE standards. They offer little or no protection and should not be used.

PERSONAL PROTECTIVE EQUIPMENT

Common types of RPE include disposable filter masks, half and full mask respirators with replacement (disposable) filter cartridges.

RPE should have a filter rating **FFP3, FFP2 or FFP1**. The filters can protect against different hazards (such as dusts or vapours) and are colour coded to show what the filter is suitable for.

- FFP3 will offer the maximum protection. These reduce the dust you breathe in by a factor of 20.
- FFP2 are suitable for medium risk. These reduce the amount of dust you breathe in by a factor of 10.
- FFP1 are suitable for light duty work. These reduce the amount of dust you breathe in by a factor of four.

e.g. **FFP3 gives a protection factor of 20, meaning that for every 20 dust particles outside of the mask it is predicted that only one would pass through the filter material.**

FFP3 masks are the most common type used in the construction industry. (FFP1 or FFP2 masks are unlikely to give enough protection against hazardous dusts.) The seal should be tight against your face. Facial hair can reduce the effectiveness of the RPE. You should be clean shaven at the start of each shift to make sure the RPE seals properly against your skin.

The risk assessment or control of substances hazardous to health (COSHH) assessment should say what type of filter is needed.

 If in any doubt – ask. You have the right not to breathe in harmful dusts and fumes.

Protect yourself today – breathe easily tomorrow.

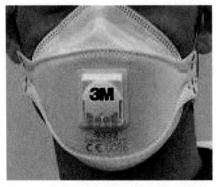

Disposable half-mask respirator. (Note the two straps passing above and below the ears)

Reusable half-mask respirator

CONTENTS

Asbestos

What your site and employer should do for you	56
What you should do for your site and employer	56
Introduction	57
Where you will find asbestos	57
Asbestos surveys	59
Working with asbestos	59
Asbestos removal	59

ASBESTOS

What your site and employer should do for you
1. Make sure that information about the location of asbestos is made available to the workforce and others who may be affected. (This information should be in the site asbestos register.)
2. Arrange for health surveillance where legislation or policy requires it.
3. Develop suitable control measures to prevent exposure to asbestos.
4. Provide you with suitable training, relevant to the level of work being carried out.
5. Provide appropriate personal protective equipment (PPE), where needed.
6. Provide supervision to make sure the control measures are being used properly.

What you should do for your site and employer
1. Follow any system of work that your employer has in place.
2. Never assume that all of the asbestos located in a building has been found.
3. Make sure that you wear any PPE properly.
4. Stop work, keep away and warn others if you think you have discovered anything thought to be asbestos or to contain asbestos.
5. Attend health surveillance, when requested by your employer.

ASBESTOS

Introduction

Asbestos is a naturally occurring mineral substance found in the earth's crust. It was mined around the world for hundreds of years, but many countries have now stopped mining and banned the use of it (including the UK) because of the health risks.

Asbestos is a harmful substance that continues to kill many people every year. It is the biggest occupational killer in the UK.

Asbestos mining

Microscopic asbestos fibres are invisible and are easily disturbed.

Breathing in any type of asbestos fibre can cause lung disease (such as asbestosis, lung cancer and mesothelioma).

Asbestosis is scarring of the lung that happens after heavy exposure to asbestos. It can cause shortness of breath and in severe cases it can be fatal.

Asbestos-related lung cancer is similar to lung cancer caused by smoking but it is caused by exposure to asbestos fibres.

Mesothelioma is a cancer of the thin, protective membrane (pleura) surrounding the lungs, heart and abdominal cavity, which is only caused by exposure to asbestos. It can take up to 50 years after exposure before symptoms develop. Mesothelioma is fatal in all cases, with death often occurring within months of diagnosis.

 Around 5,000 people in the UK die each year as a result of past exposure to asbestos.

Simon's Story:
Living with an asbestos related disease

Where you will find asbestos

The identification of any substance thought to be or to contain asbestos can only be checked by laboratory analysis.

There are three main types of asbestos.

Blue asbestos (crocidolite). Has good insulation properties and was often used in lagging. It is highly carcinogenic.

Brown asbestos (amosite). Good for insulation, particularly in asbestos insulation board (AIB). It is also highly carcinogenic.

ASBESTOS

White asbestos (chrysotile). The most common form of asbestos. It is commonly found in asbestos cement products. It is also a known carcinogen.

 Carcinogens are substances that can lead to cancer (cancer-causing agents).

It has been estimated that asbestos was included in over 5,000 construction products from the 1940s to the 1980s. Asbestos and asbestos containing materials (ACMs) were mainly used in the following.

- Sprayed coatings to ceilings, columns and beams.
- Lagging of boilers, pipework or ducting.
- Pipework and boiler gaskets.
- Suspended ceiling tiles and floor tiles.
- Partitions and ceilings.
- Soffit panels and window boards.
- Asbestos cement roofing sheets, water tanks and pipes.
- Asbestos cement board or sheets used as permanent formwork.

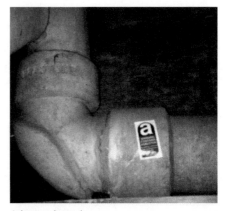
Asbestos pipework

 Any building constructed before 2000 must, by law, be considered as potentially containing asbestos, and must be subject to an appropriate survey.

Asbestos is often hidden above suspended ceilings and behind walls. Workers should stay aware and assume that asbestos may still be present, even after a survey has been completed.

If you think that you have discovered or disturbed asbestos, you must make sure of the following.

- Stop work immediately.
- Warn others nearby to keep away.
- Tell your supervisor or employer.

ASBESTOS

Asbestos surveys

An asbestos survey should be undertaken by a competent person before the work starts. The results of the survey are essential in planning the work. There are two types of asbestos survey.

1. **Management survey.** To assess how to manage and protect asbestos in an occupied building.
2. **Refurbishment or demolition survey.** This must be undertaken before any invasive construction work starts. This survey is much more comprehensive than a management survey.

Asbestos can only be correctly identified by getting a sample analysed in a laboratory.

 In the absence of a laboratory report, it must be assumed that the material contains asbestos.

Working with asbestos

Any worker who is due to work in a building or anywhere on site that is likely to contain asbestos, and whose work may expose them to, or bring them into contact with, asbestos, must complete a suitable asbestos awareness course before starting work.

Workers may come into contact with asbestos when undertaking maintenance, refurbishment or demolition work, such as the following.

- Modifying the structure (partial demolition).
- Stripping out.
- Installing new plumbing, electrics, windows and soffits, loft insulation, kitchens and bathrooms.

Never attempt to work on asbestos unless you are competent and equipped to do so and, if necessary, licensed.

All work with asbestos needs a written assessment of the risks, appropriate levels of precaution (such as the use of personal protective equipment) and appropriate training.

Asbestos warning label

Asbestos removal

Projects can be categorised in the following ways.

- Licensed asbestos work.
- Notifiable non-licensed work.
- Non-licensed work.

ASBESTOS

The removal of high-risk asbestos containing materials (such as sprayed asbestos, asbestos lagging and asbestos insulating board) must be carried out by a licensed contractor.

The removal of low-risk asbestos containing materials (such as asbestos cement products and vinyl floor tiles) would not normally need a licence, as long as the correct precautions are taken (such as wearing the correct personal protective equipment (PPE) and respiratory protective equipment (RPE)). Training must be given to all employees undertaking this work.

Damaged asbestos lagging

Labelled encapsulated asbestos

If you are carrying out work with asbestos you must make sure of the following.

- You understand and follow the written plan of work.
- You use the correct RPE and PPE.
- You follow company procedures for the cleaning, maintenance and storage of your RPE and PPE.
- You maintain your RPE and PPE in a clean and efficient state, good order and repair.
- You use the washing and changing facilities provided.

You must be given awareness training if you are at risk of coming into contact with or disturbing asbestos whilst carrying out your normal everyday work.

CONTENTS

Dust and fumes (Respiratory hazards)

What your site and employer should do for you	62
What you should do for your site and employer	62
Introduction	63
Respiratory hazards	64
Controlling exposure to hazards	67

DUST AND FUMES (RESPIRATORY HAZARDS)

What your site and employer should do for you

1. Make sure that respiratory (breathing) hazards (such as dust, fumes and vapours) are either removed completely or minimised.
2. Provide equipment and systems to make sure that exposures to respiratory hazards are reduced to the lowest level possible.
3. Provide information, instruction and training on the hazards and controls (such as on-tool extraction and wet cutting methods).
4. Issue you with the correct personal protective equipment (PPE) and the correct respiratory protective equipment (RPE) needed to protect your health, and train you how to properly check it, wear it, clean it, store it and look after it, and tell you what to do if it is damaged or lost.

What you should do for your site and employer

1. Make sure you know what respiratory hazards are and how they can damage your long-term health.
2. Use any equipment provided to make the task safer (such as on-tool extraction methods).
3. Follow the safe system of work given for the tasks you are carrying out.
4. Look after and wear your RPE properly, and replace it, including any disposable filters, as necessary.
5. Ask, if in any doubt, and report any problems with protective equipment and any changes in your health.

DUST AND FUMES (RESPIRATORY HAZARDS)

Introduction

Respiratory hazards are substances in the air (such as construction dust) that can be inhaled and can lead to a range of illnesses and diseases. Damage is mainly caused to the lungs and airways, and includes lung cancer and silicosis. However, some respiratory hazards can also result in diseases in other parts of the body, such as the kidneys and liver.

The Health and Safety Executive (HSE) estimates that around 450 people every year are dying as a result of past exposure to silica dust, which causes lung cancer. Many workers are exposed to silica during tasks such as cutting paving slabs, blocks or kerbs.

As well as deaths from respiratory hazards, many more workers suffer chronic, debilitating health conditions (such as occupational asthma) from breathing in hazardous substances on site. Many people are forced to leave the industry because of work-related ill health.

The effects of exposure to most respiratory hazards are not always immediate. You may go home from work feeling more or less the same each day, not realising that you have been exposed to a substance that may cause problems in the future.

If you are repeatedly exposed to small doses of dust, fumes and vapours they may start to damage your body. The effects of these small doses build up over time. It can take months or even years before symptoms of exposure start to show and become a problem and by then irreversible damage to your health may have occurred.

At any one time there are far more people off work through occupational ill health than because of work-related accidents.

Exposure to everyday hazards, including things like wood dusts, flux, welding fumes, dusts from slab cutting, asbestos fibres and many more common workplace materials, can cause ill health.

Some dusts and fumes, known as **respiratory sensitisers**, can cause workers to develop an allergic reaction (such as asthma) in the airways if inhaled. When the dust or fume (allergen) is breathed in, it triggers the reaction, which can cause wheezing, coughing and breathlessness (asthma attack).

Breathing in hazardous airborne substances can also cause serious health problems, such as bronchitis and many other respiratory diseases, including various types of cancer.

Exposure to solvent vapours can cause acute (immediate) effects as well as chronic (long-term) illness. This means an affected person can suffer headaches, dizziness, unconsciousness and even death if they are exposed to high concentrations of airborne hazardous substances, particularly if exposed in a restricted, poorly ventilated or confined space.

Workers in the construction industry are exposed to many materials and products, some of which are particular respiratory hazards (such as asbestos, cement, stone, silica, lead, medium density fibreboard (MDF), plastics, epoxies and solvents).

DUST AND FUMES (RESPIRATORY HAZARDS)

Respiratory hazards

The following categories describe the main respiratory hazards that you may come into contact with on site.

Dust

Dust is produced when solid materials are broken down into finer particles. Dust may be classed as either inhalable dusts (larger dust particles that can enter the airway and may be coughed up) or respirable dusts (that go deep into the lungs and remain there). Generally, the finer the particles are, the more hazardous they are, since they can get deeper into the lungs. The most hazardous dust particles are invisible to the naked eye and stay suspended in the air for long periods.

Nuisance dust has no particular harmful properties but, if inhaled in large enough quantities, can still be harmful and cause breathlessness, wheezing, coughing and irritation.

Operative working with dust suppression to reduce airborne dust

Asbestos fibres in air are responsible for several forms of cancer, including lung cancer and mesothelioma, as well as asbestosis. The microscopic fibres are easily released into the air by disturbing asbestos or asbestos containing materials (ACMs). Asbestos is not easily recognised in materials and is often missed on construction sites. If you suspect you have found asbestos, stop work and tell your supervisor. *(For further information refer to Chapter B06 Asbestos.)*

Silica dust. Respirable crystalline silica (RCS) are microscopic particles of silica found in many construction materials (such as paving slabs, blocks and kerbs). Being exposed to silica dust for a long time can cause silicosis, lung cancer and other life-changing or life-shortening conditions. Fine silica dust is created when materials are sanded, cut or drilled.

Lead is a toxic metal that can cause many ill health effects (such as mental disturbance, effects on the digestive system, and even coma). Breathing in lead dust is only one of the ways that lead can enter the body. It can also enter by inhaling fumes from vapourised lead, by absorption through the skin from handling lead, and ingestion, via hand to mouth contact, from lead particles on hands when eating or smoking. Lead can be found in many materials used in construction, such as old paint, flashings, roofing materials and some old water and gas pipes.

DUST AND FUMES (RESPIRATORY HAZARDS)

Wood dust, created by cutting and sanding, is particularly hazardous, causing a range of health effects, from mild irritation (coughing) to sensitisation and certain types of cancer. Softwood dust is known to cause respiratory sensitisation (a form of allergic reaction) and asthma. Hardwood dust is a cancer-causing agent (a carcinogen) known to be responsible for lung cancer and cancer of the nasal passages. Medium density fibreboard (MDF) is made from separated softwood fibres, hardwood fibres and glues. It can form a particularly fine dust when cut, which increases the likelihood of dangerous amounts being breathed in.

Bird and bat droppings can be common on construction sites, particularly on demolition or redevelopment projects, and often contain a number of bacterial and fungal biological hazards. If disturbed, they can release a hazardous airborne dust that can result in severe respiratory illness (such as psittacosis) that causes flu-like symptoms (fever, headache and muscle aches) but can lead to severe pneumonia and non-respiratory health problems. If work is to be carried out in an area where pigeons, common gulls, starlings or bats have been gathering or nesting, the area must be thoroughly decontaminated first, and those carrying out the decontamination must wear suitable PPE and RPE.

> Dust is by far the most common hazard on construction sites and in construction work.
>
> Silica dust is the biggest risk to construction workers after asbestos. Heavy and prolonged exposure to respirable crystalline silica can cause lung cancer and other serious respiratory diseases.

Fumes

Fumes are produced by the heating of metals to high temperatures (such as during welding and gas cutting). Fumes are effectively gases that contain tiny metal particles (molecules) that can be inhaled. Metal fume fever is one of the main illnesses caused by breathing in metal fumes. It is caused by inhaling fumes from the welding or hot working of a number of metals, particularly galvanised steel. It is an acute illness, causing flu-like symptoms. There is scientific evidence that exposure to all welding fumes, including mild steel welding fumes, can also cause lung cancer.

Vehicle exhaust emissions

Vehicle exhaust emissions contain many hazardous substances, including toxic gases and particulates.

Carbon monoxide (CO) may cause death quickly, particularly where engines are used or their exhaust gases are discharged into confined spaces where high concentrations of carbon monoxide gas can build up quickly. Sooty particles from diesel engines are very fine, and can penetrate deep into the lung, causing lung cancer.

Personal carbon monoxide (CO) alarm

DUST AND FUMES (RESPIRATORY HAZARDS)

Avoiding the use of engine-powered equipment (such as disc cutters and generators) in enclosed spaces, spaces with restricted or poor ventilation and confined spaces will reduce the risk. However, where this cannot be avoided, adequate ventilation must be provided and carbon monoxide gas detectors should be used.

Gases

Gases are chemicals in their gaseous state at room temperature. They mix with the air we breathe, and may be a simple asphyxiant (such as nitrogen) or toxic (such as hydrogen sulphide).

Hydrogen sulphide (H_2S) is given off by rotting organic substances (such as sewage and rotting vegetation). It can build up in poorly ventilated or confined spaces (such as sewers, drains, silos and slurry tanks). Hydrogen sulphide smells of rotten eggs and attacks the nerve endings in the nose, disabling them. People think that, if they can no longer smell it, the gas has disappeared, but it is still there and has simply disabled the sense of smell. At some concentrations hydrogen sulphide can cause unconsciousness within a few breaths, and death in a short space of time.

Other toxic gases that are often found in construction are carbon dioxide (CO_2) and carbon monoxide, although they usually only reach toxic concentrations when released into confined spaces or restricted spaces with poor ventilation.

 Carbon dioxide and carbon monoxide are colourless, odourless and tasteless. Both can cause loss of consciousness and death in a short time. Acute exposure to high concentrations of hydrogen sulphide can result in collapse (knockdown), respiratory paralysis, cyanosis, convulsions, coma, cardiac arrhythmias and death within minutes.

Vapours

Vapours are the gaseous state of substances that are liquids at room temperature. (As fumes can be said to be gases carrying suspended solid particles, vapours can be said to be gases carrying suspended liquid particles.)

Vapours usually form when a substance evaporates (for example, the vapour from glue or paint). Solvents are also used for many different construction processes and can release high levels of vapour.

Vapours can build up quickly in poorly ventilated or confined spaces and can cause headaches, dizziness, collapse and, in rare cases, death. Many vapours are also highly flammable and can be explosive. Long-term exposure to some solvent vapours can also cause chronic health effects (such as liver and lung damage).

Gases and vapours can build up quickly in confined spaces

DUST AND FUMES (RESPIRATORY HAZARDS)

Mists and aerosols

Mists and aerosols are fine droplets of liquid suspended in air. The droplets are often of respirable size, which means that, if inhaled, they can get deep into the lungs. Some examples of hazardous mists and aerosols are listed below.

Twin pack paints, used for creating high-quality finishes, usually contain isocyanates, which are powerful respiratory sensitisers, as well as being respiratory irritants. They can cause acute symptoms (such as respiratory irritation and bronchitis), but they can also cause asthma if inhaled over a period of time.

Legionella are common bacteria that live in water and watercourses (the aquatic environment). The bacteria are dormant in cold water but thrive and breed in warm, still and stagnant water. If you are working with water in a spray or mist form, any legionella bacteria that it contains can easily be breathed in, resulting in legionellosis (a group of diseases caused by legionella bacteria).

The most commonly known of these diseases is probably legionnaires' disease, which is a potentially fatal form of pneumonia. In construction, water aerosols may arise from dust suppression systems, vehicle washes, hot water storage tanks and sprinkler systems (where the water may be laying still in pipes or tanks for some time).

Controlling exposure to hazards

The following simple steps should be taken to protect you from respiratory hazards.

Avoid creating dust

Choosing the right equipment or method of work can protect workers' lungs and potentially remove the risk altogether. Pre-order sized materials rather than cutting them on site, or use a block splitter rather than a disc cutter as this creates less dust and is quicker.

Stop the dust getting into the air

If the creation of dust cannot be eliminated then minimising the dust being released should be the priority. This can be done by capturing the dust or dampening down.

Capturing the dust. Some materials (such as wood) do not suit the use of water to suppress dusts, so dust extraction should be considered. When your employer purchases or hires tools they should make sure that they have on-tool dust capture (extraction) wherever possible. Cleaning dust from work areas and tools is far better if a vacuum is used rather than sweeping with a brush.

On-tool dust extraction

DUST AND FUMES (RESPIRATORY HAZARDS)

Dampening down or wet cutting. This is the cheapest and most effective way of minimising exposure. Water helps to form a slurry that prevents the majority of dust becoming airborne. Not only does it reduce what is breathed in, but it also has the benefit of less cleaning up afterwards. It is important to keep the flow constant while wet cutting or grinding and to dampen down before clearing up.

Petrol driven cut-off saw with water fed dust suppression

Wear respiratory protection

Even the best control measures won't prevent dust completely, so RPE should always be worn, even if you are using dust capture equipment (extraction) or wet cutting.

> **Refer to Chapter B05 Personal protective equipment for further information about RPE.**

CONTENTS
Noise and vibration

What your site and employer should do for you	70
What you should do for your site and employer	70
Introduction to noise	71
Employer responsibilities	71
Hearing protection	72
Noise action levels	73
Introduction to vibration	74
Equipment that can cause hand-arm vibration syndrome	75
Managing vibration risks	76
Reducing vibration risks	76

NOISE AND VIBRATION

What your site and employer should do for you

1. Complete assessments to make sure that you are not exposed to levels of noise or vibration that can cause damage to your health.
2. Develop suitable control measures for managing and limiting your exposure to noise and vibration.
3. Provide you with suitable training so you understand the hazards from noise and vibration and the harm that can be caused to you.
4. Provide suitable hearing protection.
5. Provide supervision to make sure the control measures are being used properly.
6. Arrange for health surveillance where your levels of exposure state this is necessary.

What you should do for your site and employer

1. Make sure you understand the damage that too much noise and vibration can cause you.
2. Wear your hearing protection when needed.
3. Make sure that you wear your hearing protection properly.
4. Look after your hearing protection and get it replaced if it is lost or damaged.
5. When using hand-held vibrating tools do not exceed any safe time limit, if one has been given.
6. Attend health surveillance, when requested by your employer (it's for your benefit).

NOISE AND VIBRATION

Introduction to noise

Noise is sometimes referred to as unwanted sound. Damage caused to your hearing by exposure to noise can be permanent. As well as hearing loss, many workers also suffer from constant ringing or buzzing noises in one or both ears. This is known as tinnitus and there is no cure for it.

Noise-induced hearing loss can build up over time. You may not notice the effects each day, but over a period of time your hearing can get worse from the noise you are exposed to at work. When hearing damage becomes significant you may notice the following.

- You start to turn the television up a bit more than you used to.
- You struggle to hear conversations or parts of words, particularly when there is background noise (such as in social situations).

There are various sources of noise on a construction site. Many of these are made by hand-held tools and can give high noise exposures because they are close to the user's ears and because of the pitch of the sound they produce.

Other sources of noise could include the following.

- Tools that you are using.
- Tools that other people are using nearby.
- Site equipment (such as generators).
- Work equipment and hand-held tools (such as excavators and concrete breakers).

Damage can be caused by continuous noise (such as drills and breakers) and peak noise, usually from impacts (such as from cartridge-operated tools or piling rigs).

As well as causing hearing damage, too much noise on a construction site can mask other noises (for example, warnings or the sound of vehicles approaching). This creates an extra hazard (such as being struck by site vehicles).

Employer responsibilities

Employers have a legal duty to make sure your hearing is not damaged at work. They must also tell you about the risks from noise and take steps to protect your hearing. They can do this in the following ways.

- By selecting and using the quietest equipment available (such as generators with silencing kits fitted).
- By planning to do the work using the quietest methods available (such as cutting holes with rotary diamond cutters rather than chain drilling with hammer drills).
- By keeping you away from the noisiest activities, wherever possible.
- By giving you personal protective equipment (PPE).

NOISE AND VIBRATION

Where the noise levels are not reduced enough using the first three methods above, your employer should provide you with hearing protection to further reduce your exposure.

 Induction, briefings, site rules and signage will tell you where you should wear hearing protection. Make sure you wear it, where and when instructed to do so.

Hearing protection

There are two main types of hearing protection, each has its own advantages and disadvantages.

Earplugs

- These can be foam plugs (disposable or reusable), which may also be mounted to a band or cord.
- They must be fitted correctly so that they work properly, otherwise they have little effect.
- They must be fitted with clean hands, otherwise dirt can get into the ears and cause infection and other problems.
- You should be shown how to fit earplugs properly.

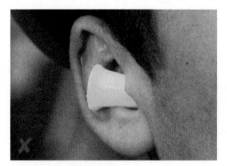
Foam earplugs worn incorrectly

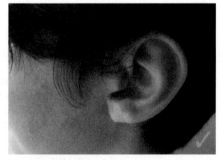
Foam earplugs worn correctly

Ear defenders (earmuffs)

- These cover the outer shell of the ear and need to make a tight seal against the head to work properly.
- They do not provide suitable protection when worn over long hair or glasses.
- They should be stored properly so they don't get dirty or damaged.
- They must be kept clean and be cleaned after use or they can become unhygienic.
- They can get hot and sweaty in hot weather and can be uncomfortable to wear for long periods.

NOISE AND VIBRATION

 Whichever type of hearing protection you use, it must be worn for all of the time that you need to wear it. Even short periods of not wearing protection have a huge effect on your daily exposure.

Noise action levels

The **daily exposure level** is the amount of noise a worker is exposed to, measured over an assumed eight-hour working day. The Noise at Work Regulations give two action levels that are the daily exposure levels at which your employer must take action. If the noise is above a certain level, the amount of time you are exposed to the noise must be reduced in line with this. The issue and wearing of PPE should always be considered as a last resort.

1. **Lower exposure action level.** At noise levels of 80 dB(A) your employer should identify the noise hazard area (by the use of signage) and make sure that appropriate hearing protection is available. These are often called advisory hearing protection zones.

2. **Upper exposure action level.** At noise levels of 85 dB(A) your employer should identify the noise hazard area (by the use of signage), provide suitable hearing protection, and make sure it is used correctly. These areas are often called mandatory or compulsory hearing protection zones (which means *everyone must* wear hearing protection).

 If you have to raise your voice to be heard by someone 2 m away from you, then the noise level is in the range of 85 dB(A). If you have to raise your voice to be heard by someone who is 1 m away, then the noise level is likely to be in the range of 90 dB(A).

A hand tool giving 95 dB(A) at your ear would result in you reaching the 80 db(A) action level in just 15 minutes, and the 85 dB(A) action level in under an hour.

A petrol cut-off saw giving 105 dB(A) at your ear would result in you reaching the 80 dB(A) action level in under two minutes, and the 85 dB(A) action level in about five minutes.

If you are exposed to the upper exposure action level your employer must put in place a hearing protection programme, which should include health surveillance.

Your employer can insist on the use of hearing protection at any time and/or at any noise level, but in doing so they must make sure that the compulsory use of hearing protection does not increase or affect other risks.

The HSE website contains an audio file that demonstrates the effects of noise-induced hearing loss

NOISE AND VIBRATION

Introduction to vibration

Exposure to vibration can have serious health effects. Vibration from hand-held tools can cause hand-arm vibration syndrome (HAVS), which affects many workers in the construction industry. Whole-body vibration (WBV) can also occur in the construction industry and may affect drivers of construction vehicles, often resulting in back injury.

HAVS is a range of conditions caused by exposure to vibration through the hands. The condition is permanent and disabling; there is no cure. The damage being done affects the nerves, blood vessels, muscles, bones and other tissues of the hands and arms. Symptoms within the hands often include the following.

- Tingling.
- Numbness.
- Loss of sensation.
- Loss of manual dexterity.
- Painful throbbing.
- Extreme burning in the fingers and hands, especially after they have been exposed to cold or have become wet.

There may also be muscle fatigue and loss of grip strength, as well as conditions such as carpal tunnel syndrome and tennis elbow (lateral epicondylitis).

A common symptom of HAVS is finger blanching, a discolouration of the fingers, which typically indicates a temporary obstruction of blood flow. This may be temporary at first but can become permanent with further exposure.

People with HAVS often have problems with everyday tasks (such as fastening buttons). It is also painful, particularly when the hands are cold or damp.

HAVS is a progressive illness. This means that if you have the condition the symptoms will not improve but will get worse over time.

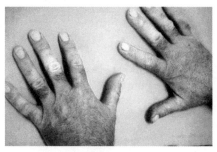

Finger blanching caused by vibration

NOISE AND VIBRATION

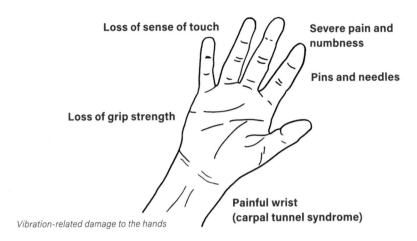

Vibration-related damage to the hands

 If you are diagnosed with HAVS you must inform your employer as they have a legal duty to report this to the Health and Safety Executive (HSE).

Equipment that can cause hand-arm vibration syndrome

HAVS is caused by the use of hand-held equipment that produces high levels of vibration. Typical equipment that produces enough vibration to cause HAVS includes the following.

- Compressed air breakers, pneumatic drills, concrete pokers and scabblers.
- Hand-held drills, including hammer drills and rock drills.
- Disc cutters, angle grinders and wall chasers.
- Sanders, circular saws and planers.
- Plate compactors and strimmers.

Your chance of getting HAVS depends on the vibration dose you receive each day. The dose is a combination of the amount of vibration produced by the tool and the length of time it is used for each day. A common way of managing the risk to workers from vibration in the construction industry is by limiting the daily dose (the amount of time workers spend using vibrating tools).

NOISE AND VIBRATION

Managing vibration risks

Your employer has a duty to make sure that you are not exposed to levels of vibration that can cause damage to your health. They must make sure of the following.

- Your risk of exposure to vibration is assessed.
- The following measures are put in place to control vibration.
  - Buy equipment that produces the lowest levels of vibration.
  - Choose work methods that avoid exposing workers to vibration, such as the use of machine mounted rather than hand-held breakers.
  - Make sure that equipment is well-maintained and attachments (such as bits and discs) are not too worn.
  - Limit the time you use vibrating equipment, so that you do not receive a potentially damaging vibration dose, and provide you with information on usage limits (trigger times).
- You are given information and training so that you know the risks from vibration.
- Health surveillance begins if you are exposed to significant levels of vibration.

Reducing vibration risks

Although your employer has a duty to protect your health, there are things that **you** can do to reduce your risk of developing HAVS.

- Be aware of the information provided on usage limits (trigger time limits) for vibrating tools, and make sure you do not exceed them.
- Relax your grip and let the tool do the work. Applying too much force or gripping too tightly results in more vibration being sent to the hands.
- Take regular breaks when using vibrating tools, and during those breaks exercise your fingers to improve the blood supply.
- Keep your hands warm – cold hands are more prone to damage from vibration.

Remember the following.

- Smokers are more likely to get HAVS than non-smokers (as smoking reduces circulation).
- Wear gloves to keep the hands warm. (The use of so-called anti-vibration gloves is **not** recommended. In most cases they have little or no effect on vibration sent to the hand, and in some cases can make matters worse.)

 The damage caused by exposure to too much vibration is permanent, debilitating, can affect your personal life and could end your career.

CONTENTS
Hazardous substances

What your site and employer should do for you	78
What you should do for your site and employer	78
Introduction	79
How hazardous substances can affect your health	79
How hazardous substances can get into your body (routes of entry)	79
What your employer should be doing	79
Identifying hazardous substances	80
Disposal of hazardous substances	81

HAZARDOUS SUBSTANCES

What your site and employer should do for you

1. Assess the risks of using substances (control of substances hazardous to health (COSHH) assessments).
2. Use a less hazardous substance where possible.
3. Make sure there are facilities to store and dispose of substances correctly.
4. Control exposure by providing engineering methods (separation, insulation, isolation and local exhaust ventilation) and the use of safe systems of work.
5. Provide information and training on the hazards, risks and controls, and on safe systems of work.
6. Monitor the work and the workplace to make sure the controls are effective.
7. Have procedures and equipment in place to deal with incidents or spills.
8. Make sure workers are trained in incident and emergency procedures.
9. Provide health surveillance for employees, where required.

What you should do for your site and employer

1. Make sure you fully understand the hazards of using the substance, before you use it.
2. Follow the safe system of work and controls, as explained in the COSHH assessment.
3. Store, dispense and dispose of the substance correctly.
4. Wear the correct personal protective equipment (PPE), as identified in the COSHH assessment.
5. If you feel ill when using any substances, stop work and tell your supervisor or site management.
6. Report any incidents or spills to your supervisor or site management.
7. Attend health surveillance sessions, where you need to do so.

HAZARDOUS SUBSTANCES

Introduction

A hazardous substance can be described in the following ways.

- Any substance (solid, liquid, gas, dust, fibre, fume, vapour, spray or mist) that could harm your health (including things such as solvents, paints, adhesives, exhaust fumes, cement, wood dusts and so on).
- Any dust, fibre or fume given off by a work process.
- Any substance that could harm the environment.

How hazardous substances can affect your health

Immediate (acute) – for example, burns to the skin from acids or chemicals.

Long-term (chronic) – for example, developing asthma from wood dust, or cancer from asbestos exposure.

Disease – for example, through contact with contaminated soil or water, or a skin sensitiser (such as cement) causing dermatitis.

Sensitising – over a period of time you start having more serious allergic reactions to lower levels of, or less exposure to, the substance, and/or a quicker reaction after exposure.

How hazardous substances can get into your body (routes of entry)

Inhalation – breathing in dust, fumes, vapours and fibres.

Absorption – contact with your skin or mucous membrane (lips, inside nose and eyelids), or open sores or wounds on your body.

Ingestion – being swallowed or eaten (usually from holding food or smoking with dirty hands).

Injection – needles, sharps or high pressure (such as water jetting), breaking the skin and forcing a substance into the body.

What your employer should be doing

Under the Control of Substances Hazardous to Health (COSHH) Regulations employers should undertake the following.

- Read and understand the manufacturer's information on the substance.
- Carry out a risk assessment on how and where it is to be used.
- Use a less hazardous substance, if available.
- Put in place measures to prevent or control any exposure.
- Make sure you know the controls.
- Monitor to make sure the controls are effective.

HAZARDOUS SUBSTANCES

 Some hazardous substances, such as those created by work processes (for example, fumes from welding or dust from cutting and sanding) do not have warning labels, as these are created by the work. Your employer has a legal duty to assess these hazards and complete a COSHH assessment in respect of the work, as part of the safe system of work.

If you are exposed to a workplace hazard your employer may ask you to attend regular health surveillance sessions, which you have a legal duty to attend.

 Health surveillance is a system of ongoing health checks to detect possible ill health effects at an early stage.

Employers can then introduce better controls to prevent any ill health effects from getting worse.

Identifying hazardous substances

The packaging or container of hazardous substances will carry one or more symbol, to indicate its hazardous properties. Some of the pictograms you will see on packaging and containers are shown in the table below.

Globally harmonised pictograms			
☠	Acute toxicity, very toxic or toxic. Can be fatal if swallowed or inhaled.	🌿	Hazardous to the environment and aquatic life.
🛢	Contains gases under pressure. May explode if heated and can cause burns.	🔥	Oxidising gases, liquids and solids. May cause or intensify fire.
❗	Harmful skin, eye or respiratory irritation. May cause an allergic reaction, drowsiness or breathing difficulties.	☣	Aspiration hazard. Damage to organs and may cause serious longer-term health hazards (such as carcinogenicity, respiratory sensitisation and reproductive toxicity).
🔥	Flammable gases, liquids, solids and aerosols. Heating may cause a fire.	🧪	Corrosive and can cause severe skin damage or burns.
💥	Explosive, self-reactive. Heating may cause an explosion.		

HAZARDOUS SUBSTANCES

 If you find a substance in a container with no label make sure that no-one else can come into contact with it, and report it to your supervisor.

Disposal of hazardous substances

Hazardous substances can contaminate land, drains, sewers, rivers and the air.

They should **never** be mixed with general (non-hazardous) waste, or be poured down drains or sinks, onto the ground, or be buried, burnt or fly tipped.

 Your site or employer should have a procedure to dispose of hazardous waste, including empty or part-used containers.

Highly flammable liquids

Highly flammable liquids (HFLs), such as thinners, solvents, petrol and adhesives, can be easily ignited, catch fire and burn fiercely.

If you have to use highly flammable liquids you must make sure of the following.

- Check there are no naked flames or other sources of ignition nearby.
- Only take the amount you need for your immediate needs.
- Always follow the storage procedures.
- Have the correct fire extinguisher at hand and make sure you are trained to use it.

 If a fire extinguisher is needed for hot-work activities it must be one specific for the work activity and the workplace. It should be issued or approved by the person authorising the hot work and you should be trained to use it. Fire extinguishers should not be taken from fire points to cover hot works, no matter how close the fire point is or how short the hot work will be.

Liquefied petroleum gas

Liquefied petroleum gas (LPG) is a form of highly flammable gases held under high pressure to liquefy them. They will readily evaporate if released from the pressure vessel (cylinder). LPG is heavier than air so will sink into excavations, basements, drains and so on.

LPG cylinders must be stored upright in a well-ventilated area and in a secure cage.

There are separate regulations covering the safe carriage of LPG cylinders in vehicles.

HAZARDOUS SUBSTANCES

LPG has a distinctive smell. If you think there might be a leak make sure of the following.

- Others are warned to evacuate the area.
- The following is carried out, only if it is safe to do so.
  - Turn the cylinder supply valve off.
  - Open doors and windows.
  - Remove any source of ignition. **Do not** allow the switching (on or off) of electrical apparatus.
- It is reported immediately.

LPG store

 Gas from a leaking LPG cylinder can expand to 250 times the cylinder's volume. It can catch fire at some distance from the original leak and flash (burn) back to the source of the leak.

CONTENTS
Manual handling

What your site and employer should do for you	84
What you should do for your site and employer	84
Introduction	85
Manual handling assessment	85
Lifting and handling	87

MANUAL HANDLING

What your site and employer should do for you

1. Reduce the need for manual handling by effective planning, delivery, off-loading and distribution of your materials and equipment, to avoid unnecessary lifting and carrying.
2. Identify work activities that pose a manual handling risk.
3. Identify the risks to your health from the manual handling of any load.
4. Ask you how you would do the manual handling task (consultation).
5. If the manual handling task has to be done provide suitable equipment and a safe method of work to minimise any risk of injury.
6. Provide any necessary equipment and training in safe manual handling.
7. Provide hazard-free access routes and adequate lighting levels.

What you should do for your site and employer

1. Follow the safe system of work.
2. Use lifting aids and equipment safely.
3. Co-operate with your employer. If in doubt, ask for advice.
4. Make sure your activities do not put yourself or others at risk.

MANUAL HANDLING

Introduction

 Manual handling **is the moving of any load by bodily force, including lifting, putting down, pushing, pulling or carrying.**

Manual handling includes the following.

- Repetitive body movements (twisting and turning).
- Horizontal movements (pushing and pulling).
- Lifting, carrying and lowering loads and materials.

Using suitable manual handling aids (such as wheelbarrows, sack trucks and brick hoists) can reduce the risk of manual handling injuries, which are the construction industry's biggest cause of ill health.

Every year an estimated 900,000 working days are lost due to handling injuries.

 Your back is strong but not as strong as you believe it is. Repeated twisting, straining and incorrect lifting techniques can, over time, lead to injury.

Manual handling injuries resulting from unsafe or incorrect manual handling can affect the following parts of your body.

- Whole body.
- Back.
- Shoulders.
- Arms.
- Hands.
- Feet and ankles.

Back injuries are most common but hernias, ruptures, sprains and strains are all conditions that can result from manual handling.

Your employer has a legal duty to avoid any manual handling activity that will harm you.

Poor posture (such as slouching on the settee, sleeping on a poor supporting mattress, and sitting in a driving position that twists or doesn't support your spine) can all add to the problem.

 Manual handling and lifting injuries are often for life.

Manual handling assessment

Where the workplace assessment shows potential risks to the health of employees from the manual handling of loads, the employer must develop a safe system of work that avoids those risks. This calls for attention to four main things.

MANUAL HANDLING

1. **Task.** What has to be achieved, by whom and when?
   - Can manual handling be avoided completely?
   - Does the task involve repetitive lifting?
   - Does the task involve repetitive twisting?
   - Can the distance a load has to be moved be reduced (for example, can it be delivered or moved nearer)?
   - Can lifting aids be used (for example, a wheelbarrow, trolley or vacuum lifters)?
   - How and from what height is the load to be lifted or lowered?
   - Will it be necessary to overreach or stretch to lift or put down the load?

2. **Individual.** Who will be involved in the task and what is their gender, build, age, experience and state of health?
   - Will they need specialist manual handling training or training in any technique needed to use the equipment?
   - Are they male or female, tall or small, young or old?
   - Do they have any existing injuries or health conditions?
   - Do they need additional personal protective equipment (PPE)?

3. **Load.** Is it too heavy to lift, can it be broken down into smaller loads, or can two people lift it?
   - Can the load be split down into smaller loads?
   - Should it be moved by two (or more) people?
   - Is the weight and centre of gravity of the load known?
   - Is the load hot, cold, wet or greasy, or does it have sharp edges?
   - Can it be gripped easily or are there adequate and suitable handholds?

4. **Environment.** What hazards are there on the route? For example, are there slippery or uneven floors, slopes, steps or narrow passages?
   - Is the floor or ground free from slip or trip hazards?
   - Does the load need to be carried up steps or stairs?
   - Are there any space constraints or obstructions that should be removed?
   - Is the level of lighting adequate?

 You can remember these four aspects by use of the acronym TILE.

Team lifting

If the load is large, heavy or awkward get help, preferably from someone of about the same size and build as you to help maintain the balance of the load during lifting. Team lifting should only be carried out with people of similar capabilities who have been trained to carry out team lifting. One person should take the lead and give clear instructions during lifting, carrying and lowering.

MANUAL HANDLING

 The only way to avoid manual handling is to use mechanical handling equipment (such as cranes, forklifts and goods hoists).

Lifting and handling

Use your body wisely (kinetic method)
- Let your leg and thigh muscles do the work.
- Try to maintain the natural curve of your spine and relax your muscles.
- Avoid flexing your back any further once you have started to lift.
- Do not snatch the load as you lift.
- Avoid leaning sideways or twisting your back, especially when your back is bent.

Good practice method for kinetic lifting

Putting the load down – plan before you start
- To floor level – you will probably need to carefully reverse the lifting method.
- To a higher level – check first, as it may be safer to set the load down and get assistance to lift it higher.
- Heavier loads should be stored at waist height.

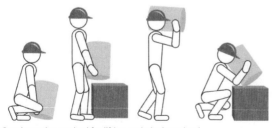

Good practice method for lifting and placing a load

MANUAL HANDLING

10

CONTENTS
Safety signs

What your site and employer should do for you	90
What you should do for your site and employer	90
Mandatory signs – must do	91
Prohibition signs – must not do	91
Warning signs	92
Emergency escape and first-aid signs – safe conditions	93
Fire-fighting signs	93

SAFETY SIGNS

What your site and employer should do for you
1. Provide signs that are suitable for the hazards.
2. Provide signs for general information (such as the location of assembly points, parking, pedestrian routes and welfare facilities).
3. Make sure that safety signs follow standard designs in respect of shape, colour and the use of pictograms.
4. Display safety signs as identified in your safe system of work.
5. Maintain signs and make sure they are clearly visible.
6. Remove signs if they are no longer needed.

What you should do for your site and employer
1. Understand what the signs mean.
2. Follow the instructions or directions on any signs.
3. Do not vandalise or remove any signs.
4. Report any damaged or missing signs.
5. Speak to your supervisor if you are confused by a sign, there seem to be too many signs or your work area needs extra signage.

SAFETY SIGNS

Mandatory signs – must do

General mandatory sign (to be accompanied where necessary by another sign)

Safety harness must be worn

Safety helmet must be worn

Protective eyewear must be worn

Safety boots or safety shoes must be worn

Safety gloves must be worn

Prohibition signs – must not do

No smoking

No naked flames

No access for pedestrians

Do not extinguish with water

Not drinkable

No mobile phones

SAFETY SIGNS

Warning signs

General warning sign
(to be accompanied where
necessary by another sign)

Combustible or flammable
material

Explosive material

Toxic material

Corrosive material

Radioactive material

Overhead load

Industrial vehicles operating

High voltage

SAFETY SIGNS

Emergency escape and first-aid signs – safe conditions

Emergency escape route signs

First-aid signs

First-aid

Emergency eyewash

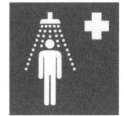
Emergency shower

Fire-fighting signs

Fire hose reel

Fire ladder

Fire-fighting equipment

Fire emergency telephone

Fire extinguisher

Fire alarm call point

SAFETY SIGNS

11

CONTENTS

Fire prevention and control

What your site and employer should do for you	96
What you should do for your site and employer	96
Introduction	97
Ignition and fuel sources	97
Emergency procedures, testing and controls	98
Hot works	98
Highly flammable liquids and gases	99
Portable fire extinguishers	99

FIRE PREVENTION AND CONTROL

What your site and employer should do for you

1. Undertake a fire risk assessment at the start of the project, review it as the project and risks on site change, and inform you of actions to take in the event of an emergency.
2. Display a fire action plan that details the actions you should take in the event of an emergency, the fire escape routes and fire assembly points.
3. Provide, maintain and test fire detection and fire-fighting equipment. Keep escape routes and fire exits clear of all obstructions and make sure they can be easily identified.
4. Have a system to control hot work, such as a hot-work permit.
5. Provide a secure storage place for gas cylinders, flammable liquids and materials.
6. Provide means to make sure rubbish and waste do not build up.
7. Provide and identify designated smoking areas, where applicable.

What you should do for your site and employer

1. Know what to do, and how to raise the alarm, if there is a fire.
2. Store materials and fuels in designated areas.
3. Keep exit routes and fire points clear at all times.
4. Practise good housekeeping and clear up your waste.
5. Know when and how to get a hot-work permit, if required.
6. Report anything you feel could be a fire risk (such as damaged electrical cables, hot lamps near combustible materials, and signs of arson).
7. Do not move fire extinguishers or signs from their designated points.
8. Do not smoke on site, unless there is a designated area where it is allowed.

FIRE PREVENTION AND CONTROL

Introduction

Fire kills and injures many people every year in the UK. On construction, demolition or refurbishment sites there can be a high risk of fire. Smoke and fire can spread rapidly and this risk can increase depending on the stage of the project.

How fire starts

For fire to start there must be three elements (the fire triangle).

1. Heat or ignition (such as a spark).
2. Fuel (something that burns).
3. Oxygen (air).

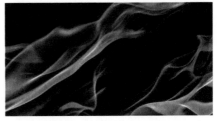

The fire triangle

Fighting fire

Fire feeds on fuel and oxygen. A fire can be put out by removing any one of the following.

- The heat or ignition source (cooling with water).
- The fuel (starving, for example, by turning off the gas supply).
- The oxygen (smothering with foam or a fire blanket).

A fire can start more easily than it can be stopped

Ignition and fuel sources

Fires need an ignition source: a form of heat that can set fire to flammable materials nearby. If rubbish is allowed to build up, or solvents and paints are left open, they will create a dangerous environment that could be set on fire by any of the following sources of ignition.

- Naked flames.
- Sparks from cutting activities.
- Overheating of power tools and electrical equipment.
- Gas or electric heaters.
- Smoking, or discarded matches or cigarettes.
- Overheating of lights (particularly halogen).

Bonfires are not allowed on site, unless the site has a permit.

FIRE PREVENTION AND CONTROL

Emergency procedures, testing and controls

Your employer will tell you the following information at the site induction.

- Where the emergency assembly points are and how to get there safely in the event of a fire.
- How to raise the alarm.
- How to call the emergency services.
- Where you can smoke on site (if smoking is allowed).
- The day and time that fire alarm tests will take place.
- Who is trained to use fire-fighting equipment.
- Who the fire wardens and marshals are.
- Who to report fire risks, signs of arson and break-ins to.

An example of a good fire point on site

 The emergency escape routes may change during the project, so you will need to regularly familiarise yourself with any new routes.

Hot works

Hot works can be any work where heat, sparks or naked flames are produced (such as welding, grinding or soldering).

If you are undertaking hot works you must work in accordance with the site rules. This often means obtaining a hot-work permit from your supervisor or site management before you start any work.

What a hot-work permit will tell you

- What you must do before you start any work, and when you can start work.
- How to prevent sparks, heat and flames from spreading.
- Which type of fire extinguisher you should have available and how many (often two will be needed).
- The site rules about maintaining a fire watch during the hot-work activity and for how long after the end of the hot works the fire watch must be maintained. (This is usually a minimum of one hour after hot work ends, but clients may set longer periods, depending on the risks involved.)
- When you must stop work towards the end of the working day to allow enough time for the fire watch to be carried out before the site is left overnight.

FIRE PREVENTION AND CONTROL

 You should be informed of fire safety and evacuation procedures during your site induction (for example, the means of raising the alarm, what the fire alarm will sound like, when testing will take place and the location of fire-fighting equipment and fire assembly points).

 If you hear the fire alarm (other than when it is a test), you must stop work and go to the fire assembly point immediately.

Highly flammable liquids and gases

Highly flammable liquids are those that can ignite at temperatures below 32°C.

When using these you must get out only the amount you need for the day and the liquids must be kept in a closed container. You should not work with flammable substances near naked flames or sparks, and spillages must be cleared up immediately.

Examples of highly flammable liquids include the following.

- Petrol.
- Thinners (for example, white spirit).
- Solvents.
- Adhesives.

Compressed gas must be kept in an upright position when being used, transported or stored.

205-litre diesel drum fitted with safe extraction equipment

Liquefied petroleum gas (LPG) should be stored in fireproof compounds or cages, which will stop vapours from building up in the event of a leak.

Compounds should have a level base, good ventilation and be surrounded by secure fencing.

 Never store LPG cylinders below ground in cellars or basements.

Portable fire extinguishers

To minimise the need to use portable fire extinguishers, it is vital that everyone on site is vigilant and any hot work is controlled. Anyone who may need to use a portable fire extinguisher should be trained and competent to do so.

FIRE PREVENTION AND CONTROL

As a guide, the table below shows the types of portable fire extinguisher and the types of fire they can be used on. Choosing the wrong fire extinguisher can increase the risk to the person operating the extinguisher, and those in the surrounding area.

Extinguishing medium	Colour of panel	Where not to use
Water: for wood, paper, textile and solid material fires	Red	**Do not** use on liquid, electrical or metal fires
Foam: for liquid fires	Cream	**Do not** use on electrical or metal fires
Powder: for liquid and electrical fires **Specialist dry powders:** for metal fires	Blue	**Do not** use on metal fires **unless** M28 or L2 text is printed on extinguisher, which means it is suitable for metal fires
Carbon dioxide: for liquid and electrical fires	Black	**Do not** use on metal, wood, paper or textile fires
Wet chemical: for wood, paper, textile, cooking oil and solid material fires	Yellow	**Do not** use on liquid, gas or electrical fires

Note: dry powder extinguishers may be provided as well as, or substituted for, water, foam or carbon dioxide extinguishers. Extinguishers used to control Class B fires (flammable liquids) will not work on Class F fires (cooking oils) because of the high temperatures produced.

Fire extinguishers should only be used on small fires (as a guide, no bigger than a wastepaper bin) or to aid escape.

 If you have to leave the location of the fire to raise the alarm, do not return – continue your own evacuation. Fires can spread quickly.

CONTENTS
Electrical safety

13

What your site and employer should do for you	102
What you should do for your site and employer	102
Introduction	103
Electrical voltages	103
Using extension leads and cables	104
Electrical hazards	104

ELECTRICAL SAFETY

What your site and employer should do for you
1. Protect you from electrocution by identifying the location or isolation of mains electric, and both underground and overhead electricity supplies.
2. Agree a safe system of work and have in place a permit-to-work system.
3. Provide safe, temporary electrics and safety lighting.
4. Make sure electrical installations are inspected, maintained and certificated.

What you should do for your site and employer
1. Only use electrical tools or equipment in accordance with the agreed safe system of work.
2. Only use tools and equipment if you have had the correct training, and after carrying out the pre-use checks.
3. Use the correct equipment and personal protective equipment (PPE) in an appropriate manner.
4. Report any damage or faults.
5. Assume that any exposed cables or wires that you are working near are live, and do not start work until a permit to work has been issued.

ELECTRICAL SAFETY

Introduction

You can't see electricity. You can't smell it. It is dangerous and it can kill.

- There is no visible way of knowing for sure if a cable or wires are live.
- All existing underground cables should be identified from surveys and their locations checked with the use of a cable avoidance tool (CAT).
- The temporary nature of site electrical distribution systems and the possibility of them being damaged are all the more reason to be careful with electricity.
- If there are exposed cables or wires near where you are working, assume they are live, stop work and report them.
- Before you start any work, you must obtain permits to work to confirm the disconnection of any live cables.

Electrical voltages

Battery power

- Battery-powered tools are the safest option.
- The severity of any electric shock will be much lower.
- There are no trailing cables.
- A lot of sites now provide secure battery-charging lockers.

110 volt – yellow

- Electrical tools on construction sites should, ideally, be battery powered or a maximum of 110 volts.
- The standard colour code for 110 volt sockets, cables and equipment is yellow.
- You would feel an electric shock from faulty 110 volt equipment but no lasting damage should be done.

230 volt – blue

- Domestic voltage or mains power is 230 volts.
- It is commonly used for generators and electrical distribution.
- The standard colour for 230 volt outdoor use sockets, cables and equipment is blue.
- If a 230 volt cable is damaged, or if you touch the wires that make up the cable, you will get a severe or even fatal electric shock.
- 230 volt tools are not allowed on most sites.

400 volt – red

- The standard colour for 400 volt sockets, cables and equipment is red.
- It is for equipment needing a lot of power (such as a tower crane).

ELECTRICAL SAFETY

 Electricity can kill. If in doubt, stop work and report any concerns to your supervisor.

Using extension leads and cables

- Extension cables and reels should be fully unwound to avoid them overheating and catching fire.
- Do not overload extension cables with too many plugs or plug adapters.
- Where possible, route cables or leads overhead. (Cables on the floor are trip hazards.)
- Use protective ramps to protect cables from being run over and prevent trip hazards if they need to be routed at floor level.
- Don't route cables across puddles or waterlogged ground.
- Get a transformer moved rather than using several leads joined together.

Electrical hazards

- There may be overhead power lines on site, which are only completely safe if isolated by a competent and authorised person.
- People have been electrocuted and killed when items (such as ladders, scaffold towers or mobile plant) accidently touch or come close to live overhead power lines. Remember, objects only need to be close to overhead lines, not actually touching them, for electricity to jump (arc) and electrocute you.
- Burns, scorch marks or burning odours indicate an electrical fault.
- Damaged leads and cables, even if it is just the outer sheathing, must be taken out of use.
- If you see smoke from a power tool, switch it off and take it out of use.
- If a fuse blows, it means there is a fault. Switch off and disconnect the equipment and report it to your manager. Do not try to do any repairs (even changing a fuse), unless you are trained, competent and authorised to do so.

Protection devices

- Residual current devices (RCDs) are life-saving devices. They work by cutting the power quickly if there is a fault, preventing electrocution.
- Fixed RCDs are installed into the consumer unit and offer protection to all sockets on the circuit.
- Portable RCDs fit between the mains socket and provide an additional socket with an RCD fitted, into which the equipment is plugged. Portable RCDs only protect the individual who is using the equipment.
- Your site rules may require you to use an RCD with any 230 volt tool.

ELECTRICAL SAFETY

- They must be kept free of moisture and dirt and be protected against vibration and mechanical damage.
- Portable RCDs should have a combined inspection and test before first use and then every month.
- Portable RCD units should have a mechanical (trip) test before each use. You do this by pressing the test/trip button.

Portable appliance testing

- Health and Safety Executive (HSE) guidance recommends that tools, cables and equipment should be tested every three months if used on site.
- If used, the portable appliance testing (PAT) label will tell you when the last safety test was carried out. It does not tell you when the next test is due.

ELECTRICAL SAFETY

13

CONTENTS
Work equipment and hand-held tools

What your site and employer should do for you	108
What you should do for your site and employer	108
Introduction	109
Safe systems of working	109
Types of hand-held tools and equipment	110

WORK EQUIPMENT AND HAND-HELD TOOLS

What your site and employer should do for you

1. Provide the right tools, equipment and personal protective equipment (PPE) for the job.
2. Provide information and training so you know how to use the tools and equipment safely.
3. Make sure tools and equipment are inspected, maintained and certificated.

What you should do for your site and employer

1. Only use tools or equipment in accordance with the agreed safe system of work.
2. Only use tools and equipment if you have had the correct training, and after carrying out the pre-use checks.
3. Use the correct equipment and PPE in an appropriate manner.
4. Report any damage or faults.

WORK EQUIPMENT AND HAND-HELD TOOLS

Introduction

Many types of power tools and hand tools are used in the construction industry and thousands of injuries happen because of misuse and abuse.

Power tools with moving parts, rotating blades and drill bits can cause serious injury in an instant, as can non-powered hand tools (such as handsaws, trimming knives and hammers).

If you use tools regularly without having an accident, there is a danger that you can overlook or forget about the hazards and risks. **Stay alert.**

Noise, vibration and dust created when using hand-held tools and equipment can pose a serious risk to your health. Vibration can lead to hand-arm vibration syndrome (HAVS). Noise can lead to hearing loss or tinnitus (constant ringing in the ears). Dust can lead to breathing difficulties or fatal lung disease. These negative health effects may not be noticeable immediately after using a tool, but they can build up and get worse over time. To protect yourself from health problems, make sure you follow the correct procedures and wear the correct personal protective equipment (PPE).

Safe systems of working

Competence

To operate powered hand-held tools and equipment, you must be competent. To help you achieve competence, your employer has a responsibility to make sure you undertake training to give you the knowledge, understanding and skills that you need to use tools safely.

- You need knowledge of the tool and the hazards associated with its use.
- You must have an understanding of the environment the tool is to be used in and its limitations.
- You must be trained and authorised to use certain tools (abrasive wheels and grinding discs, for example).
- You need to first use the tool under supervision to build up your experience with it.

Before using any tool or equipment

- Make sure it is the right tool for the job.
- Carry out a pre-use inspection and checks.
- Make sure it has been maintained and is safe to use and, if required, has had a valid portable appliance test (PAT) carried out (usually identified by a sticker on the tool or equipment).
- If you are unsure about any of these points – ask your supervisor.

 The correct tools, equipment and PPE needed should be listed in the method statement and/or safe system of work for the task.

WORK EQUIPMENT AND HAND-HELD TOOLS

Guards and safety features

- Make sure all guards and other safety features (such as emergency stops) are in place and working properly.
- Make sure all guards and other safety features are properly adjusted to minimise any gap between the work piece and any moving part.
- Always use tools and equipment properly, and in line with your training and the manufacturer's instructions.
- Always unplug the tools from the socket or isolate equipment from the power source when changing drill bits and blades, or when adjusting guards on power tools.
- Never carry out any makeshift repairs or modifications to guarding systems. This includes electrical repairs (for example, changing a fuse) unless you are trained, competent and authorised to do so.

Protect yourself

- Do not use electrical equipment in wet or damp conditions. Air or fuel powered equipment should be used.
- Keep hands away from moving parts.
- Avoid loose clothing. (Baggy sleeves and cords hanging from hoods or jackets have caused several serious injuries.)
- Make sure power leads don't get entangled in moving parts.
- Wear eye protection suitable for the task to be carried out and the tool being used.
- You may need to wear suitable hearing protection, identified in the risk assessment, due to the noise level of most power tools.
- You may need to wear suitable respiratory protection (such as a half-mask respirator), identified in the risk assessment, due to the dust created by many power-tool activities.
- Make sure that you do not exceed the daily safe vibration exposure limits when using vibrating tools. Information about the vibration level of a tool should be included in the method statement and/or safe system of work for the task.

 Refer to Chapter B08 Noise and vibration for further information.

Types of hand-held tools and equipment

Abrasive wheels and diamond blades

- The following are common types of machine that use abrasive wheels or diamond blades.
    - Petrol cut off saws (disc cutters).
    - Angle grinders.
    - Tile cutters.
    - Masonry bench saws.
    - Wall chasers and floor saws.

WORK EQUIPMENT AND HAND-HELD TOOLS

Abrasive wheels and diamond blades can burst, shatter or suffer high speed segment loss (fragmentation) in the following situations.

- If the blade or wheel is not suitable or compatible with the power tool.
- If the blade or wheel is not fitted correctly.
- If the power tool is not used correctly.

You must be trained, competent and authorised to change or mount a wheel or blade.

When using cutting equipment, follow the safety rules listed below.

- The speed of the wheel or blade must match the machine speed.
- The wheel or blade must be suitable for the material you are cutting.

Wear the appropriate PPE and make sure the guards are properly set

- Always follow the manufacturers' inspection, maintenance and fault-finding guides.
- Avoid using cutting equipment in confined spaces, where dust and fumes create a risk to safety and health. Use dust suppression wherever possible to reduce and/or collect dust created during the cutting activity and wear appropriate respiratory protective equipment (RPE).
- You must wear high-impact eye protection when you are using abrasive wheels, disc cutters, cut-off saws or diamond blade cutters. Failure, shattering or fragmentation of a disc or blade can result in serious personal injury.

 Using a concrete diamond blade on tarmac can seriously weaken the blade. The result is that the blade will fragment (disintegrate) and nail-sized segments can fly off at high speed (approximately 180 mph).

Cartridge-operated tools

- These work like a gun by firing an explosive charge.
- They are used to fire fixings into solid surfaces (such as concrete or steel columns).
- They are dangerous, especially in untrained hands.
- You must be fully trained and authorised to use a cartridge-operated tool.

Poor practice and lack of concentration can lead to injuries

WORK EQUIPMENT AND HAND-HELD TOOLS

- Suitable high-impact eye protection and hearing protection must be worn.
- An exclusion zone must be in place during fixing operations.
- Cartridge-operated tools need to be inspected, cleaned and lubricated regularly.

Compressed gas tools (nail guns)

Nail guns are similar to cartridge-operated tools but they use a compressor, gas canister or battery as the power source.

- They should only be used by trained, competent and authorised persons.
- High-impact eye protection must be worn. Eye protection must be replaced if it is scratched or damaged.
- Disconnect the air pipe and remove the battery or fuel cell and remaining nails before clearing a blockage or cleaning.
- They need to be inspected, cleaned and lubricated regularly.
- Fuel cells or canisters must be disposed of correctly and not in general waste skips.
- Tool manufacturers often offer free, on-site training.

Chainsaws

- Chainsaws can inflict serious injuries, including amputation.
- The main hazard is that they have a fully exposed cutting chain with no protective guard.
- Chainsaws can kick back (uncontrollably kick upwards towards the operator with the chain running).
- You must be fully trained, competent and authorised to use a chainsaw, and be wearing full chainsaw protective body clothing, head protection and other PPE mentioned above.

Compressed air-powered tools

- Compressed air tools are attached to a compressor using air hoses.
- Tools include heavy duty breakers, soil picks, concrete scabblers and pokers.
- High pressure air hoses can cause serious injury if they break away from the compressor or tool. **Whip checks** should be used on every hose joint to prevent this.
- Always check hose fittings are tight and secure and whip checks are in place before use.

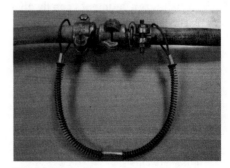

Whip checks should be used on hose joints

WORK EQUIPMENT AND HAND-HELD TOOLS

Refuelling petrol-driven hand tools

- Petrol must be kept in approved containers, and never stored where the work is being carried out.

- Refuelling should only take place at a designated, authorised refuelling point. You will be told about this at site induction.

- Refuelling areas should have impermeable floor surfaces to avoid pollution, and be well-lit and well-ventilated. Containers should be fitted with safe dispensing nozzles, otherwise a funnel should be used.

- Do not refuel when the engine is running or when any parts are still hot.

- Do not smoke while you are refuelling, or at the refuelling point.

Exhaust fumes are toxic and must not be allowed to build up in enclosed, poorly ventilated or confined spaces. Do not use petrol, diesel or liquefied petroleum gas (LPG) powered tools in a confined space or area with poor ventilation unless you have specific authority and a written safe system of work for the task. The safe system of work should include forced air ventilation and continuous gas monitoring.

Electrical hand tools

Ideally, all electrical hand tools used on site should be battery operated or 110 volts.

Before using tools with electrical leads you should carry out a visual inspection.

Check the power lead. Check for signs of damage where it enters the tool. Look to see if the cable's internal wires are visible. Are there signs of burning (overheating)? Has the cable been pulled or used to carry the tool, or is the outer insulation of the cable damaged?

Check the plug. Is it damaged in any way? Does the plug show signs of burning? Can you see the internal wires of the cable? Is there a valid PAT label attached?

Check the socket. Is it damaged in any way? Does it show signs of burning? Is a residual current device (RCD) fitted? If an RCD is not fitted, test and fit a portable RCD unit before connecting the plug to the socket.

Before making any adjustments to electrical equipment, you must switch off, disconnect from the power supply and secure the plug so that the power cannot be reconnected by accident whilst adjustments are being made.

WORK EQUIPMENT AND HAND-HELD TOOLS

Non-powered hand tools
- These may seem low risk, but they are responsible for many injuries.
- They need to be well maintained and regularly inspected.
- Well-used chisels and bolsters can form 'mushroom heads'. When they are struck, fragments can fly into the air and into the eye.
- Loose handles, sharp and blunt blades and worn parts all pose a hazard.

Lasers
- If used correctly, lasers should not pose a health hazard.
- A rotating laser means it is difficult to look directly at the beam for more than an instant.
- Static lasers (such as pipe lasers) are more of a risk.
- Exclusion zones and warning signs must be in place if high-powered lasers are being used.

 Always follow the manufacturer's instructions when using any tool or equipment.

Unmanned aerial vehicles (Drones)
Small unmanned aerial vehicles (UAVs) (also called drones) are becoming more widely available for construction activities.

Drones are being used to carry out surveys at height, in inaccessible areas and in dangerous environments, removing the risks for workers.

It is likely that drones will become commonplace on construction sites in the near future. If drones are operating on your site, do not interfere with their flight or distract the operator.

CONTENTS
Mobile work equipment

15

What your site and employer should do for you	116
What you should do for your site and employer	116
Introduction	117
Mobile plant and site vehicles	117
Management of mobile plant	117
Accidents	118
Working safely	120

MOBILE WORK EQUIPMENT

What your site and employer should do for you

1. Provide safe methods for deliveries, unloading and parking for mobile work equipment.
2. Provide suitable and well maintained mobile work equipment.
3. Provide separate routes for pedestrians and vehicles.
4. Provide signage, road markings, barriers and lighting.
5. Impose speed limits, arrange one-way systems where possible, and ban or control reversing.
6. Explain the site traffic rules at your induction.

What you should do for your site and employer

1. Follow all signs and speed limits and use unloading and parking areas as instructed.
2. Only use designated vehicle routes when operating plant and only use pedestrian routes when walking.
3. Do not bring any vehicle onto site unless your supervisor has authorised you to do so.
4. Do not move or operate mobile work equipment unless you are trained and have been authorised to do so.
5. If you are an operator, make sure all pre-start checks and daily maintenance actions are carried out.
6. Always wear your high-visibility clothing.
7. Report any mobile equipment or plant movement that you think is unsafe or too close to your area of work.

MOBILE WORK EQUIPMENT

Introduction

Mobile-operated plant and vehicles cause many accidents and serious injuries on sites every year. The accidents involve not only the operator but people carrying out maintenance to the vehicle, working close by or walking past.

Mobile plant and site vehicles

The term *mobile plant* covers all site vehicles that can move either under their own power or by being towed. Some examples are listed below.

- Dumpers.
- Excavators.
- Telehandlers and forklifts.
- Mobile cranes and piling rigs.
- Mobile elevating work platforms (MEWPs).
- HGVs, lorries and delivery wagons.
- Vans and cars.
- Road rollers, including pedestrian-operated rollers.

The term *operator* is used to describe anyone driving or controlling mobile plant.

The term *pedestrian* means anyone on foot.

Management of mobile plant

Whether you are an operator or not, the site rules for the safe operation and segregation of mobile plant and people should be explained to you during your site induction.

A well-managed site will be organised to reduce the chance of accidents between mobile plant and people on foot. The following measures should be in place.

- Separate site entrances for mobile plant and pedestrians.
- Separate routes for mobile plant and people on foot, with barriers between them.
- Mobile plant is only operated by competent, authorised persons.
- One-way systems and site speed limits.
- Amber flashing beacons on mobile plant (and, increasingly, green flashing lights indicating that seat belts are being used).
- All lights are working and switched on after dark or where natural light levels are low.
- All visual aids (such as mirrors and CCTV) are clean and in good condition.
- Turning areas are in place, so that reversing is banned or minimised.
- A vehicle marshaller to control movement of mobile plant.

MOBILE WORK EQUIPMENT

Accidents

The most common types of accident are listed below.

- Being struck by reversing or moving mobile plant.
- Loss of control or overturning when working, travelling across or manoeuvring on slopes.
- People falling when climbing into or out of the machine.
- Accidental operation of mobile plant that has been left with the engine running – often when operators are getting into or out of the machine.
- Being crushed or trapped between a structure and mobile plant as it moves or slews around.

Many accidents involving mobile plant happen because plant is large and the operator has a restricted view. Extra mirrors and CCTV are sometimes fitted to improve the operator's all-round vision. You should not rely on operators using CCTV and mirrors because, for example, a camera lens can be blocked by dirt which would affect the operator's ability to see you. **Always be aware of your surroundings and keep yourself safe.** Do not rely on the operator seeing you.

If you are close to moving or operating mobile plant you could be at risk. Whenever possible, stay within the designated pedestrian routes.

Even when not travelling, a mobile crane slewing, an excavator digging or a lorry tipping material can still be a danger if you get too close.

The chance of an accident involving mobile plant and people on foot increases after dark. Even when wearing high-visibility clothing, or if the lighting is good, the operator may still not see you. The use of floodlights and the vehicle lights should improve safety, but it is not a guarantee.

When light is poor, and after dark, you will be harder to see.

Operator's field of vision is restricted

The movement of vehicles and plant should be directed and controlled by a competent person in situations where people could be at risk, such as when a lorry is reversing or when a crane is carrying out a lift.

A vehicle marshaller controls the movement of vehicles. A signaller controls the movement of a crane or a load being lifted.

If you are the operator and you lose sight of the person directing you, you must stop and locate them before continuing.

MOBILE WORK EQUIPMENT

 Do not try to operate any item of plant unless you are trained, competent and authorised. Some companies need the operator to hold the appropriate industry competence card (such as the Construction Plant Competence Scheme).

People have been seriously injured or killed when trying to pass too close to moving or reversing mobile plant.

Plant safe zones

The following diagrams are for guidance and provide information on the safe zones applicable to a range of plant machinery likely to be used on site.

 Always signal the plant operator and wait for a positive response before entering Zone 1.

Keep out of Zone 2 at all times.

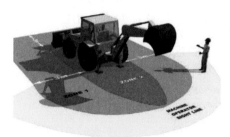 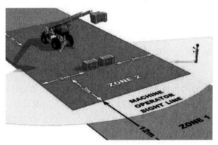

(Reproduced from 'The Delivery Hub health, safety and environment – Raising the bar 1 – Plant and equipment' from the Highways Agency, under licence of the Controller of Her Majesty's Stationery Office.)

Slewing plant

The diagram to the right shows how, as the rear of a slewing crane or excavator turns, the gap between the rear of the machine and a fixed object (such as a wall, stack of materials or other plant) becomes much smaller (becoming a **crush zone**).

This type of accident can and has happened because people on foot did not stay clear of mobile plant, took a short cut or followed a route that was not safe.

If the gap is 600 mm or less during slewing then the gap should be fenced or blocked off.

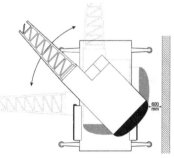

Clearance must take into account the reduced space if the crane tips

MOBILE WORK EQUIPMENT

If you are operating plant, such as a pedestrian roller or mobile elevating work platform (MEWP), always be aware of your immediate environment and what is behind or above – you may accidentally create your own crush zone.

 Stay out of the crush zone. The shortest route may not be the safest route.

Working safely

- Always wear your high-visibility clothing and keep it clean.
- Keep to the set walkways or pedestrian routes – never remove a barrier to cross a vehicle route.
- If you cannot avoid passing close to mobile plant, or if you need to speak to the operator, you will have to be patient and wait in a safe place until:
  - the operator knows that you are there, the plant stops operating and you are signalled to go past
  - the job has finished and the plant has moved away altogether.
- Stay out of plant compounds and other parking areas unless you are authorised to be there. Be alert to plant starting up and moving off, and keep out of its way.
- Do not ask for or accept rides on plant that is not designed to carry passengers. Deaths have been caused by unauthorised passengers clinging onto an item of plant then losing their grip and falling under the wheels or tracks.
- Tell your supervisor or employer about any aspects of plant operations that you think are hazardous. For example, where mobile plant:
  - uses routes that are only intended for pedestrians
  - ignores one-way systems
  - travels or operates too fast and is a danger to people
  - is operating too close to pedestrians or site operatives carrying out other tasks
  - is operating too close to excavations or at angles that could lead to overturning
  - looks to be defective.

Sometimes there isn't time to tell your supervisor or employer about a problem. If it is safe to do so, make the operator and others in the area aware of the immediate danger. Your employer will still need to know about the problem. They can then make sure that the same situation does not happen again.

CONTENTS

Lifting operations and equipment

What your site and employer should do for you	122
What you should do for your site and employer	122
Introduction	123
Examples of lifting equipment and accessories	124
Planning a lifting operation	124
Inspections	125
Thorough examination	125
Hand signals for signallers	126

LIFTING OPERATIONS AND EQUIPMENT

What your site and employer should do for you

1. Use only competent and authorised people to plan, supervise and carry out lifting operations.
2. Provide lift plans and safe systems of work, and identify exclusion zones for all lifting operations and make sure this information is communicated to you and that it is understood.
3. Make sure that lifting equipment and accessories are suitable for the task, inspected, maintained and examined.
4. Make sure that lifting equipment and accessories are marked with the safe working load or the rated capacity limits.
5. Give you information and training so you can carry out the lifting operation.

What you should do for your site and employer

1. Follow any lift plan and safe systems of work issued. Stop work and tell your supervisor if the lift plan cannot be followed.
2. Only start work if you understand your role in the lifting operation.
3. Carry out pre-use checks on equipment and report any defects before starting work.
4. Do not start work until defective items have been replaced or properly repaired.
5. Only act as a slinger/signaller if you are trained, competent and authorised to do so.
6. Attend all briefings as instructed by your supervisor.
7. Do not enter any exclusion zones unless you are authorised to do so.
8. Store lifting equipment and accessories safely when they are not in use.

LIFTING OPERATIONS AND EQUIPMENT

Introduction

Your employer must ensure the following.

- The lifting operation is properly planned by a competent person (often referred to as the appointed person, whose role is to develop or approve the lifting plan).
- The lifting operation is supervised by a competent person.
- The lifting operation is carried out in a safe manner.

Poorly planned lifting operations can go wrong, and this may result in the following.

- Deaths or injuries to people involved in the lifting operation or to people either working in the area where the lift is taking place or under the arcs of movement of the crane.
- Collapse, damage or structural failure to the lifting equipment.
- Overturning of the lifting equipment.

Accidents involving cranes and other lifting equipment are often caused by one or more of the following.

- Using the wrong type of lifting equipment or equipment with the wrong rated capacity or safe working load.
- Using lifting equipment incorrectly (unsafe technique).
- Failing to follow the agreed lift plan (system of work) for the task in hand.
- Poor assessment of ground conditions, overhead obstructions or changing weather conditions, such as high winds or lightning storms.
- Lack of co-ordination of crane movements when more than one crane is operating, and their arcs of movement overlap.
- Lack of training of the crane operators and signallers.
- Incorrect slinging of the load.
- Lifting a load of an unknown and underestimated weight or unknown centre of gravity.
- Poor inspection and maintenance of the equipment and accessories.

Lifting a generator using a lorry loader

LIFTING OPERATIONS AND EQUIPMENT

Examples of lifting equipment and accessories

Lifting equipment is any work equipment (such as cranes (mobile and static), hoists, telehandlers and excavators) that is used for lifting or lowering any load, including people.

- Scaffold hoists and gin wheels.
- Passenger/goods hoists.
- Telehandlers.
- Excavators being used as a crane.
- Rough terrain forklifts.
- Crawler, mobile and tower cranes.
- Mobile elevating work platforms (MEWPs).
- Lorry loaders.

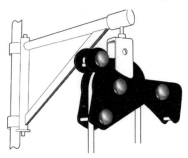

Inertia braked gin wheels (where the wheel brakes automatically if the rope is released) are now common practice and far safer than traditional gin wheels

Regardless of the type of lifting equipment being used, a lift plan must always be developed. The detail of the plan will reflect the complexity of the operation. Work involving people at height or loads which are difficult to sling is more complicated and would require a more detailed lifting plan.

Lifting accessories are items of equipment used for attaching the load to the lifting equipment.

- Chains.
- Ropes.
- Slings.
- Hooks.
- Shackles.
- Spreader-beams.
- Eye bolts.

Planning a lifting operation

All lifting operations **must** be properly planned by a competent person (sometimes called the *appointed person*) who will decide on how best to complete a lifting operation safely. This information will be recorded in a lift plan as part of the safe system of work.

The lift plan will need to consider the following (this list is not exhaustive).

- Weight of the load to be lifted.
- Ground conditions and the surrounding environment (such as waterways, railways and highways).
- Overhead hazards (for example, a power line).
- Height of the lift required.
- Type of lifting equipment to be used.

LIFTING OPERATIONS AND EQUIPMENT

- Selection of lifting accessories (for example, chains and shackles).
- Lifting team (for example, crane operators and slinger/signallers).
- Level of supervision required.
- Means of communication (for example, hand signals or two-way radios).

The lift plan will give information about any exclusion zones that are required to ensure the safety of people on site and off site. You must not enter these zones unless you are authorised to do so by the lift supervisor.

 If you are involved in any lifting operation you must be briefed on the contents of the lift plan before work starts.

Inspections

A competent person must inspect all lifting equipment and accessories at suitable intervals – before first and every use, weekly and monthly. The frequency and type of inspection will depend on the potential for the item to fail. The inspections should include visual and functional checks.

Thorough examination

As well as the inspections mentioned above, it is a legal requirement that lifting equipment should be subject to thorough examination at various intervals. Lifting equipment should be examined at least every 12 months and equipment used for lifting people should be examined at least every six months. The thorough examination must be carried out by a competent person. This will normally be a qualified engineer and is a detailed inspection that produces a report on completion. The report should be kept on file at the same location as the equipment or accessory.

Lifting equipment must be thoroughly examined as outlined below.

- When it is first used (unless bought brand new).
- If it is installed, after installation but before first use.
- If it is assembled, after assembly but before first use.

A MEWP should be thoroughly examined at least every six months

- At intervals not exceeding 12 months (six months for equipment used for lifting people).

LIFTING OPERATIONS AND EQUIPMENT

 Lifting equipment used for lifting people, and all lifting accessories, must be thoroughly examined every six months.

Hand signals for signallers

The following images are industry-recognised hand signals for people directing the movement of lifting equipment.

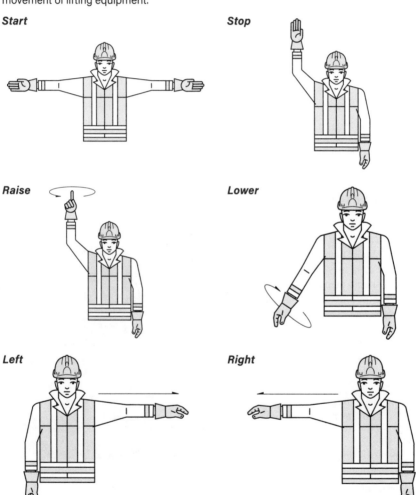

Start

Stop

Raise

Lower

Left

Right

 You must not signal plant operators unless you are trained and authorised to do so.

LIFTING OPERATIONS AND EQUIPMENT

Horizontal distance

Vertical distance

Move backwards

Move forwards

Danger

End

 The signaller should stand in a safe position, where they can see the load and can be seen clearly by the lifting equipment operator. They should face the operator if possible. Each signal should be distinct and clear. These signals have been reproduced from Schedule 1 of Health and Safety (Signs and Signals) Regulations.

(This contains public sector information licensed under the Open Government Licence v2.0.)

LIFTING OPERATIONS AND EQUIPMENT

16

CONTENTS
Working at height

What your site and employer should do for you	130
What you should do for your site and employer	130
Introduction	131
Planning the work	131
Hierarchy for working at height	132
Preventing falls	133
Fragile roofs	134
Types of access equipment	136
Fall arrest	142

WORKING AT HEIGHT

What your site and employer should do for you

1. Avoid the need to work at height if possible.
2. Provide the correct work at height equipment and make sure it is inspected and maintained.
3. Provide a collective system to prevent falls (such as a scaffold, mobile elevating work platform (MEWP) or mobile tower).
4. Minimise any risk to you if there is a chance you could fall (such as by providing safety nets, safe landing platforms or, as a last resort, a safety harness and a suitable lanyard).
5. Give you information, instruction and training so you can work at height safely.
6. Make sure that all work at height is carried out in accordance with a safe system of work and that you are briefed fully.

What you should do for your site and employer

1. Follow the agreed safe system of work and ask questions if you do not understand.
2. Use only equipment and methods you have been trained in.
3. Make sure you carry out pre-use checks on your equipment. Report any defects to your supervisor immediately.
4. Not misuse any equipment.
5. Not take risks or shortcuts.
6. Stop and seek advice if anything changes or seems unsafe.

WORKING AT HEIGHT

Introduction

- About half of all fatalities on construction sites are as a result of a fall from height.
- An average of seven people die each year as a result of falling through fragile roofs.
- Over half of all reported specified injuries are due to either falls from height or a slip, trip or fall on the same level.
- Selecting the wrong type of work equipment results in many falls from height every year.

 Many construction workers have suffered a life-changing injury as a result of a fall from height. Such an event impacts not only the worker but their family, lifestyle and ability to earn an income. It is important that all work at height activities are adequately planned, risk assessed and the correct equipment selected and used to reduce the risk of injury.

 What can you fall from?

Many falls are from poorly secured ladders, faulty or poorly used stepladders or makeshift working platforms, which offer little or no fall protection.

About seven workers a year die after falling through fragile roofs. Others suffer serious injuries and disabilities as a result of falling.

For all work at height, measures must be taken to prevent the risk of any fall that could cause injury.

Work at height is work at any height, above or below ground level, where a person could fall and be injured. It also includes instances such as working next to, or in, an open excavation, because of the risk of falling in, or of something falling in on you if you are working in the excavation.

Planning the work

- A risk assessment must be completed for all work at height.
- The following should be taken into account.
    - The complexity of the work being done.
    - Who is doing the work and for how long.
    - Weather conditions and surface conditions (such as a wet, sloping roof).
    - How to raise and store materials and equipment.
- Proper precautions and controls must be in place.

WORKING AT HEIGHT

- Choose the right equipment for the work to be done.
- People working at height must have the right mix of skills, knowledge, training and experience (competence), and the work must be adequately resourced and supervised.
- The workforce must be told about the risks involved and the risk control measures to be followed.
- Equipment must be regularly inspected and maintained and defective equipment taken out of service.
- Emergency arrangements must be in place, including properly developed rescue plans, people trained in rescue procedures and suitable rescue equipment.

Hierarchy for working at height

Any person planning work at height should always follow a hierarchy of control and consider options at the top of the hierarchy before moving down.

Step 1. Avoid working at height
e.g. assemble on the ground and lift into position using a crane or by fixing guard-rails to structural steelwork on the ground before lifting and fixing at height

Step 2. Prevent falls from occurring
Use an existing safe place of work
e.g. parapet walls, defined access points, a flat roof with existing edge protection

Step 3. Prevent falls by providing *collective* protection
e.g. scaffolding, edge protection, handrails, podium steps, mobile towers, MEWPs

Step 4. Prevent falls by providing *personal* protection
e.g. using a work restraint (travel restriction) system that prevents a worker getting into a fall position

Step 5. Minimise the distance and/or consequences of a fall using *collective* protection
e.g. safety netting, airbags or soft-landing systems

Step 6. Minimise the distance and/or consequences of a fall using *personal* protection (The last resort)
e.g. industrial rope access (working on a building façade) or a fall-arrest system (safety harness and fall-arrest lanyard), using a suitable, high-level anchor point

WORKING AT HEIGHT

Preventing falls

If work at height cannot be avoided, the best way to prevent people falling is by using physical barriers and equipment.

- Use scaffolding, mobile access towers, mobile elevating work platforms (MEWPs), podium steps or other proprietary edge protection systems.
- The minimum height of the top guard-rail must be 950 mm above the working platform.
- Gaps between guard-rails and toe-boards on the working platform must not be greater than 470 mm.
- Plastic barriers, netting or rope and pins are not suitable as edge protection as they are not rigid enough to prevent people from falling, unless they have been specifically designed for that purpose.

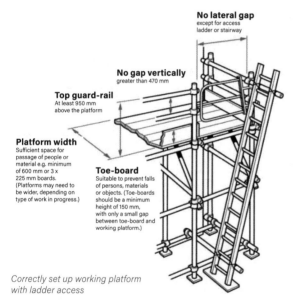

Correctly set up working platform with ladder access

Requirements for working at height

- Safe access (such as a tower staircase or secured ladder) must be provided.
- Edge protection (such as temporary guard-rails) must be put in place.
- Remember that surfaces can become slippery in wet or frosty conditions.
- If you are working on or near to a leading edge then measures (such as a physical barrier or safety net) should be installed. A harness and fall-arrest lanyard must only be used as a **last resort**.
- **Never** work on any structure where there is no protection from falls.

WORKING AT HEIGHT

Voids and holes

All voids and holes (such as those listed below) where a person could fall any distance and injure themselves must be protected.

- Floor and roof openings.
- Gaps between floor joists and roof trusses.
- Lift shaft openings.
- Service holes and risers.
- Open inspection chambers and other voids.
- Openings created during demolition.
- Open excavations.
- Openings created during site surveys.

All openings must be protected with secure barriers, covers, gates or doors, which are secured in position and display the relevant warning signs.

- Never remove a protective cover unless authorised to do so and you are protected against falling whilst the cover is not in place.
- If a hole, gap or void needs covering, you should report this to your supervisor straight away.

Properly constructed void protection, complete with proper signage

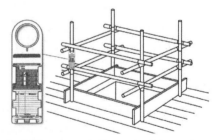

Scaffolding edge protection constructed around a void, complete with a scaffold tagging system

Fragile roofs

There are many fatal and serious injuries from people falling through fragile roofs and roof lights.

The following surfaces are likely to be fragile.

- Roof lights.
- Old liner panels on built-up sheeted roofs.
- Asbestos and fibre cement sheets.
- Metal sheets.
- Glass, including wired glass.
- Chipboard and plywood boards.
- Slates and tiles.

WORKING AT HEIGHT

Asbestos and fibre cement sheets are obvious, but fragile roof lights, which look like the more secure surrounding roof structure over time, are not so obvious.

Many people are not aware that parts of a roof are fragile.

 Never assume that a roof will support your weight and do not try to walk near the underlying supports.

All fragile roofs should be considered to be unsafe unless additional protection has been provided. It is sometimes difficult to see a fragile surface, especially if a roof is dirty, weathered, covered by moss or has been painted. Buildings with a fragile material roof covering should show warning signage (*as shown in the image below*) but this will not always be the case.

A safe system of work **must** be put in place and, as a priority, work on a roof (particularly a fragile roof) should be avoided where possible. An example of a safe system of work would be to work from underneath the roof using a MEWP.

Fragile roof and crawling board signs   *Using a MEWP to work underneath the roof*

A safe system of work should also consider the following.

- Suitable access is provided (such as a stair tower or ladder) or a MEWP is used.
- If the work has to be carried out from above the fragile surface, specialist working platforms with handrails should be provided. These are designed to span the roof purlins and evenly distribute the loading applied.
- Physical barriers or covers around or on fragile surfaces (such as roof lights or sky lights).
- The installation of safety netting, crash decks or airbags underneath the roof.
- Anchor points, work restraint systems, fall-arrest lanyards and safety harnesses.
- Arrangements for emergency situations, including how rescue will be carried out for casualties who have fallen into a safety net or are suspended by a safety harness.

WORKING AT HEIGHT

 Never try to access or cross a fragile roof without a safe system of work.

Types of access equipment

Ladders

- Working from ladders should be considered as a last resort.
- They should only be considered for light work of short duration (less than 30 minutes is recommended) that does not involve carrying equipment or materials up or down, pulling or pushing motions, or the use of pressure, and only where the use of other work equipment is not suitable.
- They are most often misused when used as a working platform.
- A risk assessment must have validated the use of a ladder to prove that it is not reasonably practicable to use other means or methods.
- It is essential that people who use ladders are trained and competent to do so.
- Always check that ladders are in good condition before use.
- Report any defects to your supervisor.
- Ladders should never be used near overhead power lines.
- Ladders should never be painted as this can hide defects or damaged parts.

If ladders are used, they **must** meet the following criteria.

- Be of the right type – Class 1 Industrial or EN 131 Trade and Industrial – heavy duty and industrial use (for professional users).
- Be in good condition.
- Be placed on firm and level ground.
- Not be at risk of being struck by vehicles.
- Use outriggers if available.
- Be set at the right length and angle for the job – **75°** or a ratio of **1:4 (1 m out to 4 m up)**.
- Be properly secured (tied at the top).
- If being used as a means of access, extend 1 m past the stepping off point.

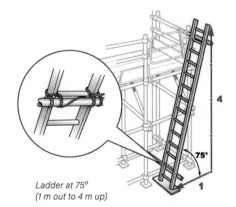

Ladder at 75°
(1 m out to 4 m up)

You should have three points of contact with the ladder at all times.

WORKING AT HEIGHT

Ladders are classified into two European-wide categories.

1. Class 1 Industrial or EN 131 Trade and Industrial – heavy duty and industrial use (for professional users).
2. EN 131 Domestic (for non-professional users). For domestic use only – should **never** be used in a construction environment.

 Class 1 Industrial or EN 131 Trade and Industrial ladders are recommended for use in construction environments as they offer the highest duty rating.

Stepladders

There are many types of stepladder on the market. Some offer good fall protection and others less so.

Stepladders should only be considered as suitable in the following circumstances.

- Where a risk assessment has shown the use of other more suitable equipment is **not** reasonably practicable.
- The work does not involve pulling or pushing motions.
- The work does not need pressure to be applied.
- The work is of a short duration, less than 30 minutes.

Wherever possible, platform steps should be the preferred option over traditional swing-back steps.

- Always check they are in good condition before use.
- Always use on firm, level ground.
- Always make sure they are fully extended and that the user faces towards the steps.
- The stepladder and you should always face the work (you should not need to twist your body to face the work, as doing so would side load the steps).
- Three points of contact should be maintained where possible.
- It is essential that those who use stepladders are trained and competent to do so.
- Report any defects to your supervisor.

Stepladder with integrated side guard-rails for additional protection

WORKING AT HEIGHT

- Never overreach or apply a side loading.
- Never stand on the top three treads of any stepladder unless it is designed to be used that way.

Podium steps

Podium steps have become a popular piece of access equipment and are a safe and efficient means of achieving low-level access if used correctly.

- They are a safe and adaptable form of access equipment, with a safe working platform if used correctly.
- They allow you to work from the working platform facing every side without it becoming unstable.
- Some podium steps are fitted with power to allow you to adjust the height easily. Some are fitted with anti-surfing features that prevent you from pulling yourself along when you are on the equipment.
- It is essential that those who build and use podium steps are trained and competent to do so.
- They can be unstable and topple over if not built or used correctly.

 You must always use the brakes when using towers and podium steps. Do not pull yourself along when you are inside it.

Podium steps in use, correctly built and wheels locked

Mobile access towers

Mobile access towers are a safe and adaptable form of access equipment if used correctly. Unfortunately, many access towers are not erected or used correctly and, as a result, become unstable and are the cause of many accidents.

You should hold a Prefabricated Access Suppliers' and Manufacturers' Association (PASMA) or equivalent qualification to erect, alter or dismantle a mobile access tower.

The training should cover how to reduce common risks associated with this type of equipment and include information on the following.

- Manufacturer's instructions for safe use.
- Positioning of the equipment – it must be positioned on suitable, firm, level ground.
- Checks made for weak areas, such as inspection chambers and lids, stop tap covers and so on.
- Making sure it is not being used close to an overhead power line where a person could make direct or indirect contact with the power line.
- Using locking wheel brakes.

WORKING AT HEIGHT

- Fitting of advanced guard-rails from a lower level to provide edge protection at the level above, without putting the operative at risk during installation.
- Checking that guard-rails and toe-boards are fitted (these must not be removed).
- Assessing loading capability (how much weight can be placed on any platform).
- Considering whether bad weather, such as strong winds, rain or snow, will cause problems.
- Assessing the chance of the tower being struck by mobile plant or other vehicles.
- Using barriers and signs to create a safety zone around the tower, to prevent other workers or the public from entering the work area.

You must make sure of the following.

- Towers are not overloaded.
- Only the internal ladder is used to access the decks.
- Working platforms are not fitted too high, so that guard-rails are too low.
- Prevent and prohibit overreaching.
- Wheels are locked when the tower is being used.

Safe working on a mobile tower, with toe-boards and guard-rails in position and wheels locked

! If you are only working on a mobile access tower and not involved in building, altering, dismantling or moving it, you should receive a toolbox talk on the risks and hazards associated with your work as a minimum. A toolbox talk will not give you the knowledge, skills, experience or authority to build, adapt, dismantle or move the tower.

Mobile elevating work platforms

- Common types of MEWP are scissor lifts and boom type (cherry pickers).
- You must only use a MEWP if you have been fully trained and are competent.
- If you are in a boom-type MEWP you must wear a full body safety harness and restraint lanyard clipped to the designated attachment point in the basket.
- Emergency ground level controls should only be used in an emergency (for example, if the operator becomes incapacitated or unwell). Someone at ground level should be trained to use the emergency controls.
- Never clip on to a nearby structure.

WORKING AT HEIGHT

- Never stand on the guard-rails of the MEWP or lean out, especially when going up or coming down.
- Never climb out of a MEWP in an elevated position unless the unit has been specifically designed for that purpose.

Scissor lift type MEWP

Cherry picker type MEWP

IPAF powered access licence

IPAF operators' safety guide

WORKING AT HEIGHT

 You should have access to the operators' manual for the machine you are using.

Scaffolding

- Scaffolding must only be built, altered or dismantled by trained and competent scaffolders.
- All working platforms must have double guard-rails and toe-boards fitted.
- Keep the scaffold working platform clean and tidy. These areas should not be used for storage unless specifically designed and designated for this purpose.
- You should always follow the safe loading information.
- Brick guards must be fitted if materials are stored above toe-board height or if there is a risk of tools or materials falling and striking someone below.
- You must never interfere with scaffolds or remove any components, no matter how simple it appears to be.
- You must never overload scaffolds.

Good example of a warning incorporated into a physical barrier – do not access scaffolding if you see this sign

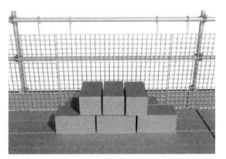

Wire brick guard with double guard-rails and toe-board

WORKING AT HEIGHT

 Employee prosecuted for dangerous work at height

A steel erection firm's employee has been sentenced at Manchester Magistrates' Court after he admitted working unsafely at height on a hotel development in central Manchester.

The Magistrates' Court heard that a member of the public had contacted the Health and Safety Executive (HSE), claiming that a man had been seen balancing on scaffold tubes in the rain while working on the roof of the multistorey hotel. HSE inspectors later found the employee working on the roof.

An investigation by the HSE found that the employee had climbed up the scaffold to hammer the steel beams into place and had not used the tower scaffold that had been made available for him. There was also a full-time scaffolder on site, available for any of the contractors to use, to make sure safe working platforms were in place.

The employee in question pleaded guilty to breaching Section 7 of the Health and Safety at Work etc. Act 1974 and was sentenced to six months' imprisonment, suspended for 18 months, he was fined £1,400 and ordered to pay costs of £2,939.18.

After the case the HSE inspector said: 'This case dealt with a serious work at height risk, which could have led to a fatal incident. The particular employee failed in his duty to protect his own safety while at work, and also placed others at risk had he dropped any tool from the position he was seen in, some 27 m above street level. During the HSE's investigation he said that he did not appreciate how high he was. Never before in my career as an HSE inspector have I seen such a staggering disregard for personal safety. It is a matter of pure luck that no-one was injured or killed. My thanks go to the member of the public who reported their concern to us, as they have been instrumental in saving a life and arguably that of anyone below him at that time'.

(Source: HSE.)

Fall arrest

Fall-arrest systems include soft-landing systems, safety netting, crash decks and safety harnesses.

- If falls cannot be prevented then the risk of injury must be minimised.
- Arresting falls by means of safety harnesses and fall-arrest lanyards must be considered as a last resort and is only acceptable when other methods (further up the work at height hierarchy of controls) have been considered and ruled out.
- The safe system of work must contain a procedure for emergency rescue.

WORKING AT HEIGHT

Safety harnesses

Fall restraint and fall-arrest harnesses are a type of personal protective equipment (PPE), and should only be used if falls cannot be prevented by physical barriers or minimised by using collective arrest systems (soft-landing systems, such as nets or airbags).

- Safety harnesses can be used to prevent a fall from occurring (fall restraint) and to minimise the consequences of a fall (fall arrest). Fall restraint is preferred to fall arrest (it is better to prevent the fall than arrest it).
- Choosing the correct type of safety harness and lanyard to be used is vital. Your employer should take into account where it is being used, how far the wearer may fall, any obstructions they may hit and any pendulum effect (swinging from side to side after the fall has been arrested).
- You must receive formal training before using a safety harness and lanyard.
- It is vital you know how to inspect a safety harness for damage, how to fit it properly, where to attach it and **where not to attach it**.

Never use a safety harness to work at height unless you have been trained.

A safety harness and lanyard could be all that prevents you from falling to your death. Make sure you clip on to a suitable anchorage point at all times.

There must be an effective rescue plan in place in case a person wearing a safety harness falls.

- They will need to be rescued quickly.
- When someone's fall is arrested and they are suspended in a safety harness they can experience a condition known as suspension syncope (where the person suspended faints and could suffer further complications).

If a person has been suspended in a safety harness, *for any length of time*, an ambulance *MUST* be called, for a professional medical assessment.

WORKING AT HEIGHT

17

CONTENTS
Excavations

What your site and employer should do for you	146
What you should do for your site and employer	146
Introduction	147
Dangers of excavations	147
Examples of good practice	148
Inspections	149
Poisonous or flammable gases and fumes	149

EXCAVATIONS

What your site and employer should do for you

1. Assess the project, identify any work that will need workers to enter or work in an excavation, and tell you.
2. Make sure there are adequate resources and equipment to prevent collapse, and a safe system of work is in place where it is necessary to enter an excavation.
3. Make sure suitable training is provided for those installing excavation and trench support systems and anyone entering or working in excavations.
4. Make sure a safe means of getting into and out of excavations (access and egress) is provided.
5. Make sure there is suitable edge protection (such as barriers and stop blocks) in place to prevent people, materials and vehicles from falling into excavations.
6. Carry out safety inspections.

What you should do for your site and employer

1. Do not enter an excavation unless a safe system of work is in place and you are trained, competent and authorised to do so.
2. Do not enter any excavation that is unsupported or has not been inspected.
3. Do not install excavation supports unless you are trained, competent and authorised to do so.
4. Do not take any risks.
5. Do not leave any excavation open or unguarded.
6. Never climb on supports or use supports or exposed services as stepping points to get into or out of an excavation.
7. Report any safety concerns to your supervisor.
8. If you feel unwell, leave the area immediately and tell your supervisor.

EXCAVATIONS

Introduction

An excavation is any hole or trench dug into the ground as part of construction or utility work. Some excavations are knee deep, but many are deeper. They do not need to be deep before becoming a serious risk or a confined space.

Every year deaths and injuries occur due to excavations collapsing, or workers being overcome by poisonous gases or striking live services.

Many accidents happen when the excavation appears to be in good condition, with no obvious hazards.

 Excavations are dangerous

A cubic metre of soil can weigh **over one tonne** (1,000 kg). **A human body cannot support that much weight.**

Even a shallow excavation can easily collapse onto you.

It can crush your legs, hips or chest and, within seconds, can prevent you from breathing.

Collapse is silent and without warning.

Dangers of excavations

The sides of a trench may look firm, but looks can be misleading. Excavations and trenches **collapse** for the following reasons.

- The sides are not supported or supports are not installed properly.
- Vehicles operate too close to the edge (for example, dumpers, excavators or other large vehicles).
- Materials and spoil are stored too close to the edge.
- The ground dries out, shrinks and collapses.
- Heavy rain weakens the ground and the sides. Groundwater in an excavation also weakens the sides.
- The excavation is too close and undermines or weakens nearby walls and structures, causing them to collapse.

 Keep vehicles a safe distance away from the edge of an excavation. A truck carrying 6 m³ of concrete weighs approximately 26 tonnes.

There are many dangers when working within excavations. To work safely the following hazards may need to be considered.

- Falls of materials, people, plant and vehicles into the excavation.

EXCAVATIONS

- Making foundations of nearby structures weak, including scaffolds and other temporary works structures.
- Accidental or deliberate contact with underground services.
- Water or other fluids entering the trench, standing water and pumping out.
- The presence of naturally occurring gases (such as hydrogen sulphide (H_2S)).
- The build up of gases that are heavier than air (such as H_2S and liquefied petroleum gas (LPG)).

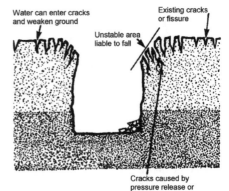

Possible dangers of trenches

Examples of good practice

Examples of good practice for excavation work are listed below.

- Avoid the need for anyone to go into an excavation.
- Install excavation supports before anyone goes into an excavation.
- Use methods that protect the person installing the support system.
- Stop vehicles from coming too close (for example, use wheel stop blocks).
- Provide a safe way to get in and out (such as a tied ladder).
- Only work within the safety of the protected area in open excavations (for example, within the confines of a drag box or trench supports).
- Provide fall prevention around the excavation (for example, a handrail or extended trench sheets).
- Put up barriers around the excavation, particularly where the public have access.

Safe working in a deep excavation using trench boxes and a tied ladder

EXCAVATIONS

- Provide lighting; excavations and the guarding around them should be lit during the hours of darkness or other periods of reduced natural light, particularly if the excavation is in a place where the public have access.

Inspections

A competent person must inspect the excavation before the start of each shift (including after breaks), or after any event (such as heavy rainfall) that may have affected its strength or stability. Something may have happened to affect the excavation's stability or something may have fallen from or into it. All inspections must be recorded and defects put right immediately.

Poisonous or flammable gases and fumes

Poisonous gases and fumes (such as those listed below) can be heavier than air and can 'pour' over the edges and start filling up an excavation or a confined space.

- Exhaust fumes from petrol or diesel-powered plant.
- Naturally occurring gases (such as methane and hydrogen sulphide), which seep out of the ground.
- Fumes or vapours from solvents (such as welding plastic pipes, epoxy resins or sealants).
- Liquefied petroleum gas (LPG).
- Carbon dioxide (CO_2) and other pipe freezing gases.

Safe systems of work may include the following.

- Using a gas detector to test the air before entry and then monitoring continuously.
- Pumping in fresh air.
- Using a solvent-free product that does not give off fumes.
- Wearing breathing apparatus as a **last resort**.

 Always be aware of gas hazards

You may not be able to see or smell gas in an excavation or a confined space.

If you are in an excavation or a confined space and feel light headed, dizzy or can smell gas: **warn others – get out – stay out – report it immediately**.

EXCAVATIONS

 Excavation collapses after dangers are ignored – Director and excavator operator jailed

A 52-year-old director of a house building company (working as the site manager at the time of the accident) has been convicted of gross negligence manslaughter following the death of a workman on a construction project in 2014. The conviction followed a nine-week trial at Northampton Crown Court into the death of a 33-year-old father-of-five whilst he was employed as a ground worker at a building site in Collyweston, Northants.

The court heard evidence of how the workman had been standing next to a deep trench, which had been incorrectly excavated by the excavator driver, when a wall of the trench collapsed, burying the workman beneath the collapsed material. Despite the efforts of fellow workers he was pronounced dead at the scene after his body was recovered. Northamptonshire Police and HSE investigators found that the sides of the trench had not been properly or adequately secured and that the site manager (the company director) and excavator driver both ignored basic safety measures.

In June 2017 the court convicted the defendants, as follows.

- The site manager and director of the company was found guilty of gross negligence manslaughter and received a four-year sentence – two years in custody and two years on license. He was also ordered to pay costs of £90,500.

- The self-employed excavator driver, who traded as a demolition contractor, was found guilty of a failure as a self-employed person to discharge his duty to ensure the health and safety of persons not in his employment, contrary to Section 3(1a) of the Health and Safety at Work etc. Act 1974. He was also found guilty of failing to take all practicable steps to prevent danger, contrary to Regulation 31 of the Construction (Design and Management) Regulations 2007. He was given a 12-month sentence – six months in custody and six months on license. He was ordered to pay £20,000 towards costs.

Speaking after the verdict a detective superintendent from Northamptonshire Police said: 'This was a tragic loss of a young life that could so easily have been avoided. The defendants were both experienced in ground works and failed to show even the most basic safety measures to prevent harm to workers, such as the young father who lost his life. He leaves behind five young children who are now being cared for by his parents. They were in court throughout the trial and have shown true dignity and strength despite listening to harrowing evidence. This has been a very lengthy and complex investigation and I would like to thank the jurors for their dedication and attention to detail during this long trial'.

CONTENTS
Underground and overhead services

19

What your site and employer should do for you	152
What you should do for your site and employer	152
Introduction	153
Underground services	153
Overhead services	154

UNDERGROUND AND OVERHEAD SERVICES

What your site and employer should do for you

1. Make sure no works are carried out in roadways or highways without the proper authority and licence.
2. Provide training for supervisors and workers to allow for works in highways and roadways (New Roads and Street Works Act (NRSWA) and Chapter 8).
3. Provide training for supervisors and workers to allow for the identification of underground services.
4. Make contact with all utility owners and receive plans and drawings.
5. Arrange for the disconnection of underground services, where possible, and obtain certificates of disconnection.
6. Where underground services are not disconnected make sure all workers know those services are live.
7. Carry out a site survey and scan for services. Locate, identify and indicate the location of all underground services before excavation starts.
8. Develop a safe system of work for performing excavation works and uncovering services. Inform, instruct and train workers in the safe system of work, including any permit to work and permit to dig procedures.
9. Make sure there is a procedure for the discovery of unidentified, underground services or objects and train workers on those procedures.
10. Where overhead cables are present, speak with the power supply company to find out if the power can be isolated, or the minimum safe working distance from which a safe system of work can be developed.
11. Put up goalposts and barriers and introduce control measures (isolators on plant) to make sure nothing and no-one goes into the exclusion zone.

What you should do for your site and employer

1. Do not work in any roadways or highways without the proper authority and licence.
2. Do not work in highways and roadways unless you are trained, competent and authorised to do so (New Roads and Street Works Act (NRSWA) and Chapter 8).
3. Do not use any scanning or survey equipment (CAT and genny) unless you are trained, competent and authorised to do so.
4. Follow the permit to work and permit to dig procedures. Do not take any risks.
5. Follow the procedure for the discovery of unidentified, underground services or objects.
6. Never climb on supports or use supports or exposed services as stepping points to get into or out of an excavation.
7. Where overhead cables are present, follow the safe system of work and follow the minimum safe working distances.
8. Do not move any barriers and never enter into the exclusion zone.
9. Report any safety concerns to your supervisor.

UNDERGROUND AND OVERHEAD SERVICES

Introduction

There are many injuries and deaths each year caused by accidental contact with underground and overhead services (such as gas pipes and electricity cables). These incidents are often due to poorly planned working activities that result in contact with underground services or overhead power lines.

Underground services

Underground services include the following.

- Electricity cables.
- Gas mains.
- Water mains.
- Sewers and drains.
- Telecommunications or fibre optics.
- Oil or fuel pipes.

Thousands of service strikes happen every year. Many result in serious injury and some are fatal. It is vital that any excavation work, no matter how big, small, deep or shallow, is properly planned.

Before any excavation or digging, every effort should be made to find existing services. These methods will show roughly where the services are.

- Refer to service drawings.
- Contact the utility company.
- Carry out a ground radar survey.
- Use cable avoidance tool detection (such as CAT and genny equipment).

A cable avoidance tool and signal generator (CAT and genny)

Hand digging trial holes to expose the services is the most accurate way of finding them.

Examples of the services you are most likely to find and their relevant colours are shown below (this list is not exhaustive).

Service	Colour
Electricity (all voltages)	Black or red
Water	Blue, black or grey
Gas	Yellow
Communications	Grey, white, green, purple or black

 The table above describes utility industry practice. However, not all apparatus will conform to the recommendations for colour coding. Some older utilities will have no marker tape or warning tiles.

UNDERGROUND AND OVERHEAD SERVICES

Safe systems of work may include the following.

- Not using forks, picks or other bladed tools near live underground services.
- Always treating services as live until confirmed otherwise.
- Never assuming a service runs in a straight line.
- Not using power tools or excavators within 500 mm of the indicated line of a service.

If you find an unexpected underground service, report it immediately.

Underground services

 If you strike a live service – if it is safe to do so, move away, do not go back, warn others and report it.

Overhead services

Every year people are killed or seriously injured when they come into contact with overhead electricity power lines.

If there is contact with a power line, or even if a piece of equipment gets too close to it, the electricity can be conducted to earth, which can cause a fire, an explosion and shock or burns to anyone touching the machine or equipment.

Overhead lines can be difficult to spot, particularly in foggy or dull conditions, but often people just fail to look up.

Overhead cables carrying electricity are generally uninsulated. You must be mindful of the following.

- Electricity will flow through any conductor that comes into contact with it (such as a metal ladder, a scaffold pole or a raised excavator bucket).
- Electricity may jump through the air (arc) to anything nearby that will conduct electricity.

The siting or use of cranes, lifting appliances or any conductor near to overhead power cables is dangerous and extreme care must be taken. If there is a danger to people with scaffold poles or other conducting objects then barriers should be used to keep people and mobile plant out of the area.

Power lines should be isolated and made **dead** or suitable precautions taken to prevent danger before any work takes place.

 Electricity travels at the speed of light – more than 186,000 miles per second. You and the people you work with don't.

UNDERGROUND AND OVERHEAD SERVICES

Always contact the power supply company before siting or using any plant or conductor near to overhead power cables. They will tell you the voltage of the supply and the minimum safe distance. If it is not possible for the power to be isolated, they will tell you the minimum safe working distance from which a safe system of work can be developed.

A safe system of work will often involve putting up barriers and possibly introducing other control measures to make sure that nothing and no-one enters into the exclusion zone.

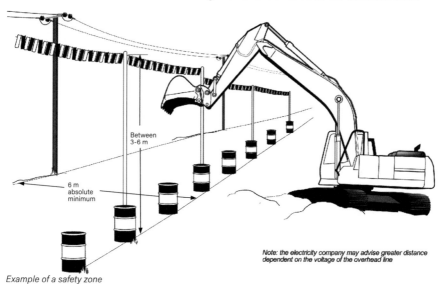

Example of a safety zone

Note: the electricity company may advise greater distance dependent on the voltage of the overhead line

⚠ Overhead power line electrocution

A construction company was prosecuted after a crane operator suffered an electric shock when the equipment he was using came into contact with overhead power lines. The sub-contractor was using the crane to move sections of steel. As they started to lift a section of steel using the crane, the hook block came into contact with an 11 kV power line and he suffered an electric shock. The sub-contractor was resuscitated but now suffers from long-term memory loss. The electricity company had warned the contractor about the presence of the overhead power cables, and had received advice on the removal of the power supplies running across the site. However, no measures were put in place by the company to prevent plant and equipment accessing the area beneath the power lines or for the power supply to be diverted or isolated. The company was fined £20,000.

The HSE said: 'This terrible incident could have been avoided had the company placed physical barriers on site so that no plant or equipment could gain access to either side and directly below the overhead power lines, or if the high-voltage cables were diverted or isolated'.

UNDERGROUND AND OVERHEAD SERVICES

19

CONTENTS
Confined spaces

What your site and employer should do for you	158
What you should do for your site and employer	158
Introduction	159
Hazards	160
Working in a confined space	161

CONFINED SPACES

What your site and employer should do for you
1. Assess the project, identify any work that will need workers to enter or work in a confined space, and tell you.
2. Assess the hazards and risks from confined space work and develop safe systems of work, with rescue plans.
3. Make sure suitable confined space training is provided for supervisors, workers and rescuers.
4. Make sure that anyone entering a confined space is properly trained and adequately resourced with rescue equipment and trained rescuers.
5. Carry out safety inspections.

What you should do for your site and employer
1. Do not enter a confined space unless a safe system of work is in place and you are trained, competent and authorised to do so.
2. Do not take any risks.
3. Do not leave any confined space open or unguarded.
4. Report any safety concerns to your supervisor.
5. If you feel unwell, leave the area immediately and tell your supervisor.

CONFINED SPACES

Introduction

Workers can become trapped or overcome by fumes, vapours, poisonous or explosive gases when working in confined spaces. This can lead to people dying.

 Many workers who tried to rescue workmates without a proper rescue plan and the necessary equipment have been overcome by toxic or suffocating gas and fumes, or a lack of oxygen, and have collapsed and died, adding to the tragedy.

What is a confined space?

Confined spaces are not just tanks, vessels or chambers. They do not have to be totally enclosed.

They can be any place where there is a risk of the following.

- Reduced or increased levels of oxygen in the air.
- The presence or build up of poisonous gases, fumes or vapours.
- The presence or build up of flammable or explosive gases or vapours.
- Drowning due to an inrush of liquid, or being engulfed or buried in free flowing solids (such as sand).

Depending upon the work hazards many places could be classed as a confined space. Some examples are listed below.

- Excavations and trenches.
- Inspection chambers, sewers and soakaways.
- Service tunnels and shafts, plant rooms and boiler rooms.
- Basements, voids and staircases.
- Lofts, roof voids and attic spaces.
- Unventilated rooms and rooms with sealed windows.
- Oil storage tanks or water tanks above or below ground.

An unventilated room may not seem like a confined space. However, if you are using a substance in the room that gives off toxic or hazardous vapours then you could be overcome and fall unconscious.

 Toxic paint remover fumes cause death of worker

A worker died after breathing in toxic fumes while carrying out restoration work in a bathroom at a flat in South West London. The worker was using an industrial paint and varnish remover to strip resin coating from the bath. The room did not have enough ventilation and the 55-year-old victim was overcome by the fumes. He died at the scene and was discovered later by the occupant of the flat.

CONFINED SPACES

Depending upon the work activity, any space can become a confined space

Entry to confined space with lifeline and rescue equipment

Hazards

A lack of oxygen

The air that we breathe contains around 21% oxygen and, at that level, people can work without difficulty. If the oxygen level falls below 10% it will cause breathing difficulties, unconsciousness and possibly death.

A reduction in oxygen can be caused by the following.

- Hot works or machinery that burns up the oxygen.
- People breathing.
- Rust (oxidisation) inside enclosed tanks.

Too much oxygen

If the air contains too much oxygen this can be a major hazard. Organic materials (such as oil and wood) will be able to catch fire and burn more easily, and ordinary materials (such as paper and clothing) will burn more fiercely.

An increase of only 4% oxygen is enough to create a hazard. This may happen by accident.

- In oxyacetylene and oxypropane processes, sometimes not all of the oxygen supplied to a cutting torch is used. Some may be released, increasing the atmospheric oxygen above the normal 21%.
- Leakage from torches or hoses may go unnoticed (such as during meal breaks or overnight). For this reason, they should be removed and properly isolated at every break time.

CONFINED SPACES

Toxic and flammable atmospheres

Oxygen can be replaced by asphyxiating, toxic or flammable gases in the following ways.

- By stirring up sludge or slurry in excavations, which may release methane or hydrogen sulphide (H_2S).
- Natural methane venting from the ground or rotting vegetation.
- Using substances that give off fumes or vapours (such as solvents, paints and resins).
- Sewage, giving off hydrogen sulphide, which smells like rotten eggs and will drive out or dilute the oxygen in the air.
- Chalky ground, which gives off carbon dioxide (CO_2), especially when wetted by acidic rain.
- People breathing out carbon dioxide.
- Gases (such as liquefied petroleum gas (LPG), methane or oxygen enrichment), which build up to form a highly flammable atmosphere.

An overturned tanker or a large spill may release petrol or dangerous chemicals into the drainage system. The vapours can travel hundreds of metres.

Hostile environments

Apart from the hazards mentioned, other dangers may arise within a confined space. Some examples are listed below.

- Difficulty using electrical and mechanical equipment in a confined space, and exposure to higher-than-usual levels of noise from the equipment.
- Extremes of heat, which can have harmful effects and may be intensified in a confined space.
- High humidity levels, which can interfere with the body's natural cooling mechanism, preventing sweat from evaporating.
- Excessive sweating, which will cause the body to lose vital salts.
- Difficulty getting into, or out of, and working in a confined space, which may involve working at height.

The potential hazard of an inrush of water, gas, sludge and so on, due to a failure of walls or barriers, or leakage from valves, flanges or blanks, must all be considered at the risk assessment stage.

Working in a confined space

Your employer should identify if the work activity, hazards and area of work is, or should be, classed as a confined space. Anyone working in a confined space must be properly trained.

Work in a confined space needs the following safety documentation to produce a safe system of work.

CONFINED SPACES

- A suitable and sufficient task and site specific risk assessment.
- A method statement, including a rescue plan.
- A permit to work (to manage and control entry).

The safe system of work will identify important issues, such as those listed below.

- Who the supervisor will be. (The supervisor should make sure the task has been properly planned and check safety at each stage. They will need to be present while work is underway.)
- Safe access and egress (how to get in and out safely).
- What tools and materials to use, and how they should be used.
- The type of air and gas monitoring equipment and alarm system.
- Who can enter and for how long (time limits).
- What personal protective equipment (PPE) and respiratory protective equipment (RPE) to use.
- The emergency arrangements.
- The rescue equipment and trained rescue team.

Emergency exit from a confined space using escape breathing apparatus (EBA)

The air within the space must be tested before entry, and constantly monitored while the space is occupied using a meter with an audible and visual alarm. If the alarm sounds, the area must be evacuated as quickly and safely as possible. There should be one person (known as the attendant) at the entrance to the confined space, whose job is to raise the alarm and start the rescue plan if things go wrong.

There must always be a clear means of communication between workers inside the space and those outside.

In some cases, it may be necessary for site management to provide information about the confined space work to the emergency services, in case they may need to be called or involved in the event of an emergency.

 Never try to rescue someone unless you are part of a properly trained and equipped rescue team. Use the time to get expert help and call the emergency services.

CONFINED SPACES

 Workers die in an open-topped inspection shaft

At Carsington Reservoir in Derbyshire, four young, physically fit men, aged between 20 and 30, died at the bottom of an open-topped inspection shaft. Naturally evolved carbon dioxide had displaced the oxygen, but no tests were made before the first man entered. He collapsed and the three other men in turn climbed down, trying to rescue their colleagues. They were overcome by the lack of oxygen and later died.

CONFINED SPACES

20

CONTENTS

Environmental awareness and waste control

What your site and employer should do for you	166
What you should do for your site and employer	166
Introduction	167
Sustainability	167
Environmental responsibilities	168
Pollution	169
Waste materials	174
Nuisance	175
Plants and wildlife	176
Archaeology and heritage	176

ENVIRONMENTAL AWARENESS AND WASTE CONTROL

What your site and employer should do for you

1. Develop and tell you the procedures for avoiding nuisance and pollution and for protecting wildlife and heritage environments.
2. Agree procedures to safely store, distribute and use materials to avoid damage and pollution.
3. Provide procedures to segregate and dispose of waste correctly.
4. Explain specific environmental issues at your site induction.
5. Provide information on environmental emergency response procedures and provide the resources (such as spill kits) for immediate action.
6. Use a colour-coding system to identify the different types of drainage on site.

What you should do for your site and employer

1. Turn off plant, equipment and taps when not in use to save energy and water and reduce nuisance.
2. Reuse materials where possible to save resources and reduce waste.
3. Where possible, avoid or control the creation of dust, chemical pollution and noise.
4. Avoid creating waste wherever possible and dispose of waste correctly.
5. Know where the spill kit is and how to use it, and report any incidents.
6. Show respect and consideration for neighbours of the site and members of the public.
7. Report any environmental concerns you may have about pollution, the protection of wildlife, historic buildings and archaeological objects.

ENVIRONMENTAL AWARENESS AND WASTE CONTROL

Introduction

Environmental protection is increasingly recognised as being one of the most important issues in the world today. To create a more sustainable built environment, the construction industry must take vital action to protect our water supplies and natural resources (such as iron, oil, copper and gas), reduce pollution that affects people's health and speeds up global warming, and avoid industrial or agricultural processes that destroy ecosystems and wildlife habitats.

The construction industry is the single biggest consumer of our natural resources in the UK, using around 420 million tonnes of materials each year. The industry also generates a huge amount of waste (around 100 million tonnes each year); about 20 million tonnes of waste goes to landfill, including 10 million tonnes of new, unused building products.

The poor management and inefficient use of materials and resources on a construction project can lead to large amounts of waste. This is unsustainable, costly, bad for the environment and it can be unsafe.

The following are some of the actions that result in the inefficient use of materials and the creation of waste.

- Poor design.
- Incorrect ordering or over-ordering.
- Incorrect storage and management of materials leading to damage.
- Poor workmanship.

Sustainability

You will increasingly hear the word *sustainability* being used in connection with construction work. Carrying out construction in a sustainable way helps everyone.

Sustainable buildings are built with materials that have a low environmental impact. They are energy efficient, reduce water consumption, create minimum waste and improve the natural environment and human health.

Sustainable construction is a responsible way to avoid environmental damage and, where possible, to improve the environment without causing problems for future generations.

Examples of good practice are listed below.

- Designing out waste, such as making the unit off site, making site dimensions to match standard product sizes, or adjusting site levels to reduce the amount of excavated material going off site.
- Using local labour and services to support the local economy and reduce energy and vehicle emissions associated with travel.
- Consider using raw materials and finished goods that are produced in the local area to minimise transport and reduce the project's carbon footprint.

ENVIRONMENTAL AWARENESS AND WASTE CONTROL

- Saving water by turning off taps, using hoses with triggers, collecting rainwater for use on site and recycling water from concrete wash-out areas.

- Saving energy and reducing nuisance by good maintenance of vehicles and turning off equipment when not in use.

- Avoid creating pollution that will damage the environment.

- Using reclaimed or recycled materials or materials with a high recyclable content.

- Only using timber products that can be fully traced to sustainably managed forests.

- Reusing leftover materials to conserve raw materials and save the energy it would take to produce new.

- Segregating waste into different types, so it can be reused or recycled more easily.

- Putting in place good storage management to avoid damage, pollution, double handling, rain damage and breakages.

Collecting rainwater for reuse

 Fuel, energy, water and materials are often wasted on construction sites – practices which need to be stopped.

Environmental responsibilities

Responsibilities of the person in charge of the site

The person in charge of the site has legal responsibilities to make sure of the following.

- The client's environmental requirements are carried out and met.

- Environmental legislation and planning conditions are met.

- A named individual is appointed with responsibility for environmental issues.

- Environmental damage and nuisance are prevented during and after construction.

- Information is provided to all site staff on activities that will affect wildlife, areas of historical and archaeological importance, and other special protection orders that may be in force on the site (such as tree preservation orders).

- Employees are told about the control of substances hazardous to health (COSHH), such as chemicals and fuels, to make sure these substances are stored and handled correctly to prevent pollution.

- Employees have the resources available and are trained how to use pollution response equipment (for example, spill kits and booms).

ENVIRONMENTAL AWARENESS AND WASTE CONTROL

- Waste materials are properly segregated to support recycling and avoid pollution.
- Waste is correctly handled and companies who transport, treat or dispose of waste are registered and licensed.
- Accurate records and documentation are kept for all activities regarding transfer, treatment and disposal of waste.

Your part in preventing environmental damage

You should be given instructions, training and advice during and after site induction so you understand the following.

- The site's environmental rules (such as how to dispose of your waste).
- What damage your work can have on the surrounding environment.
- What procedures and controls are in place to avoid creating waste, nuisance and pollution.
- What you need to do as an individual.
- What to do in an emergency (for example, correct use of a spill kit).
- How, and to whom, to report any environmental concerns or issues.
- That you should stop work and seek advice if you uncover archaeological objects, disturb wildlife or damage protected historic buildings or monuments.

Bats roosting

Stonehenge

Pollution

Causes of pollution

- Deliberate or accidental leakage of substances (such as cement, silt, grout, sewage, chemicals, oils/greases or vehicle fuels) that soak into and contaminate the ground, rivers, streams or ditches.
- Disposing of contaminated liquids into drainage systems, both foul and surface water.
- Deliberately or accidentally allowing waste material to escape from the site and contaminate nearby land and/or watercourses, via leakage or being blown by the wind.

ENVIRONMENTAL AWARENESS AND WASTE CONTROL

- Mixing contaminated, hazardous materials with other waste (for example, putting rags used to clean up an oil spillage in with general waste materials).
- Mixing plasterboard or gypsum-based waste with biodegradable waste.
- Allowing smoke, fumes or dust to contaminate the air and for residue to settle on surrounding surfaces.
- Causing too much noise, light or vibration, which can affect the quality of life of people who live or work nearby and can also have a negative impact on wildlife.

Why pollution occurs

- Poor storage of fuels, chemicals and waste.
- Not having properly protected, bunded (containment area that captures any leaks, preventing pollution) storage areas to contain any leaks or spills of harmful liquids (for example, oils, fuels and solvents).
- Not protecting or covering stockpiles of soils and aggregates (which could be contaminated with hazardous substances), resulting in them being blown by the wind, creating dust, or being washed into the ground or watercourses by rainwater.
- Failing to adequately protect waste skips, resulting in rainwater washing harmful residues out.
- Rain and muddy surface water running off site onto roads, into drains and watercourses.
- Not having processes and equipment in place to monitor, report and control environmental incidents.
- The poor maintenance of plant and equipment resulting in fuel and other leaks, excessive noise and/or air pollution.
- The illegal burning of waste materials, and fly tipping.

 A spill involving just one litre of oil can contaminate one million litres of drinking water.

Your part in preventing pollution

To avoid creating pollution you should always do the following.

- Follow the instructions in the control of substances hazardous to health (COSHH) assessment and any site rules when using any substance, particularly with regard to storage and disposal.
- Keep the lids on tins of paints, adhesives and solvents when not in use.
- Keep oils, fuels and chemicals within bunded areas when not in use.
- Prevent spillages by careful handling and pouring of harmful liquids.
- Keep harmful substances at least 10 m away from watercourses, drains and ditches.

ENVIRONMENTAL AWARENESS AND WASTE CONTROL

 Pollution spreads easily

Spilt or leaking oils and fuel can be particularly damaging to the environment. It is possible for leaking substances to soak deep into the ground, damaging the nutrients within the soil that help plants and crops to grow, and polluting groundwater, which in some cases becomes domestic drinking water.

Five litres of oil can completely cover a lake **the size of two football pitches**. If it gets into surface water or **groundwater**, pollution may appear **several miles away** and could poison fish or other living organisms.

 Spilt materials being cleaned up must be properly contained using absorbent materials, not washed down with detergent.

Refuelling site vehicles or construction equipment must be carried out by an authorised person. Where possible the refuelling should take place in a controlled area with a hard surface that prevents spilt fuel from soaking into the ground. If refuelling has to be carried out away from these controlled areas, where the ground is unprotected, make sure you follow the points below.

- A drip tray or absorbent mat must be used (it must be cleared of any spilt fuel afterwards by using absorbent spill clean-up materials).
- Refuelling must take place at least 10 m from watercourses or drains. (Where 10 m cannot be achieved, speak with your supervisor who should then put in place further control measures.)

Pollution-absorbent mats, known as plant nappies, capture fuel and oil leaks, preventing ground pollution

 Regulatory bodies recommend that refuelling is always carried out under supervision.

Where possible, refuelling should be carried out using a pumped supply through a nozzle fitted with an automatic cut-off to avoid over-filling and spillages. Where necessary, funnels should be used to help prevent spills. Never leave a tank to fill unsupervised and always replace lids and caps on containers.

ENVIRONMENTAL AWARENESS AND WASTE CONTROL

Do not dispose of harmful substances into drains or gullies. To identify the different types of drain on site, drain covers and gullies should be colour coded as shown in the table below.

Blue	Surface water (such as clean, uncontaminated rainwater).
Red	Foul water (such as sewage and silty run-off water).
Red 'C'	Combined surface and foul water (both of the above descriptions).

You should not assume that drains marked in red are suitable for the disposal of contaminated water or chemicals. Local Authorities or water authorities must be told and a permit received before any hazardous fluid is put into the drainage system. Hazardous fluid should not be disposed of in drains marked in blue.

If the product you are using displays this sign on the label or COSHH assessment then it is harmful to the environment. Any substances left over, including the container, must be disposed of in line with the COSHH information, regulations and site rules.

Concrete and cement washout is highly alkaline and can cause severe pollution. Water from washing out any mortar or concrete mixing plant or ready-mix concrete lorries must not be allowed to flow into any drain, watercourse or to ground. It should, where possible, be recycled for further washout processes.

Hazardous to the environment and aquatic life

Waste concrete and washout water can be put in a lined skip, allowing settlement of solids. Once the solids have settled and if the appropriate environmental consents (permissions granted by Environment Agencies) are in place, the water can be pumped to a foul sewer or taken away by tanker. To boost sustainability, the settled solids can be recycled into the works.

Washout areas should be designated and at least 10 m away from watercourses and drains.

If you are involved in pumping water out of an excavation (de-watering) you must be aware that silty water must be treated before it can be discharged into surface water drains, rivers, streams or ditches. Site management or your supervisor must make the decision whether the water is silty and needs treatment prior to discharge.

A lot of construction work takes place on what are known as *brownfield sites*. This means that the land has been built on before. If, in its earlier use, the land was used for industrial purposes, it may contain hazardous substances. In most cases pre-construction surveys will have been carried out to identify any hazardous materials and your employer should include this information as part of your site induction. However, you must immediately **stop** and tell your supervisor if you find either soil that has a strange smell and/or appears to be oily, or fragments or clumps of fibres that could be asbestos or other hazardous materials.

ENVIRONMENTAL AWARENESS AND WASTE CONTROL

Pollution incidents

If you are involved in a pollution incident (such as a chemical spill or discharge of contaminated water), make an initial assessment of the risk. Only take action if it is safe to do so and you have the appropriate resources, such as a spill kit and the correct personal protective equipment (PPE). If your assessment of the risk places you in danger, raise the alarm and inform your supervisor. **Do not put yourself at risk.**

If you are aware that an environmental incident or spill has occurred then act quickly and follow these four steps: stop – contain – notify – clean up.

1. **Stop** further contamination with corrective actions (such as standing up oil drums or closing a leaking valve).

2. **Contain** the contamination by using spill kits, or build earth or sand bunds to prevent the contamination from spreading. If you cannot contain the spill make sure you direct flows away from unprotected drains.

3. **Notify** your supervisor or employer of an incident as soon as possible. Do not ignore or cover up the incident over fears of being blamed, as **not reporting an incident can lead to far greater consequences**.

4. **Clean up** the spilt material and any contaminated materials and dispose of them appropriately.

As well as the four-step process above, the following actions should be taken.

Eliminate all sources of ignition in the immediate area. Do not switch plant or equipment on. Remove sources of heat and sparks.

Cover drains or inspection chambers to stop the substance entering unprotected drainage systems.

Check the spill has not reached any nearby drains, inspection chambers, watercourses, ditches, ponds or other sensitive areas.

Dispose of all contaminated materials used to contain the spill (such as absorbent granules, soil or cleaning cloths) in the correct hazardous waste skip.

Spill kit deployed

Report the facts to your supervisor or employer, who will notify the appropriate Environment Agency and take action to prevent the incident happening again.

ENVIRONMENTAL AWARENESS AND WASTE CONTROL

Waste materials

Your employer has a responsibility to identify how and where waste is created. In line with the waste hierarchy, they should identify opportunities for waste prevention, reuse, recycling and reducing the amount of waste going to landfill.

Always check if someone else can use what you are about to throw away, as it might have a use elsewhere.

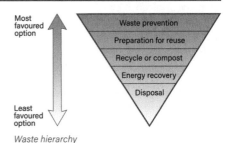

Waste hierarchy

Different types of waste should be segregated into different skips or bins so that they can be recycled more easily. Recycling waste means that it does not have to go to landfill.

To help, colour-coded labels are often put on skips or bins to show what type of waste should be put in them.

 Hazardous waste must never be mixed in with other types of waste.

Segregated waste streams

If hazardous waste gets into landfill it can be damaging to the environment. It must be removed from site by specialist contractors and treated before being recycled or disposed of correctly.

Examples of hazardous waste

- Asbestos.
- Batteries.
- Used spill kits.
- Fluorescent light tubes.
- Waste solvents (such as white spirit, oil and bitumen-based paints).
- Epoxy resins and mastics.

Oily rags segregated into hazardous waste containers

Environmental damage can result from poor waste management in the following circumstances.

- Poor housekeeping creating an untidy site, exposing unprotected waste to the weather and scavenging animals.
- Where controls of waste segregations are insufficient.

ENVIRONMENTAL AWARENESS AND WASTE CONTROL

- Where hazardous waste is mixed with other waste (such as asbestos cement mixed with rubble, or difficult waste such as plasterboard being mixed with biodegradable material).
- Insufficient checks to avoid using unregistered waste carriers who illegally dispose of or fly tip the waste.
- Inappropriate disposal of contaminated concrete washout water into surface water (such as streams and rivers).

Everyone has a part to play in preventing environmental damage. By being aware of the potential environmental risks that arise from your work, you will be able to avoid incidents, which will help to create a more sustainable and healthy environment for everyone.

Nuisance

No-one on site, or in neighbouring locations, should have to suffer nuisance and possibly ill health because of site activities.

Talk to your supervisor if you think that the work you are doing might be causing a nuisance to other people. This may be the case if your work activities result in any of the following nuisances occurring.

- Creating delays and congestion on local roads because of poor traffic management planning and logistics.
- Contractors and site operatives parking cars without giving consideration to local residents.
- Causing excessive noise, particularly during unsocial hours.
- Creating vibration that can be felt off site.
- Mud or dirt being deposited on public roads or footpaths.
- Vehicle fumes and noise causing a nuisance to nearby properties or people.
- Generating foul smells, smoke and dust.
- Creating off-site dust clouds from haul roads.
- Site lighting or task lighting shining onto nearby properties, people or road traffic.

Managing dust and emissions

Many contractors sign up to the Considerate Constructors Scheme, which requires them to maintain high standards of respect and consideration for site neighbours and the general public.

ENVIRONMENTAL AWARENESS AND WASTE CONTROL

Plants and wildlife

Landscapes, plants and wildlife are valued by the public and are a vital part of maintaining a healthy environment. Damaging, disturbing or removing plants or animals can speed up rates of decline. For protected species, this decline can be rapid and companies or individuals found guilty of contributing to their decline will face legal action resulting in prosecution and fines.

Construction activities can negatively affect plants and wildlife in the following ways.

- Removing or damaging habitats through vegetation clearance.
- Disturbing aquatic wildlife and impacting water quality.
- Disturbing wildlife through noise and vibration.
- Damaging trees and hedgerows.
- Spreading invasive plant species, such as Japanese knotweed, through inappropriate disposal.

Your employer should inform you about, and you should follow, all of the protection measures that are in place to protect plants and wildlife.

Archaeology and heritage

Archaeological remains and historic buildings provide a valuable historic record and identity of the country, adding to a sense of belonging for the population. They are valued as an irreplaceable part of the country's national heritage and are therefore given protection against unlawful or accidental damage. They also support the economy through tourism.

For these reasons, your employer should tell you about the controls and your responsibilities when working on or near historic buildings or archaeological and heritage sites.

These controls could include the following.

- Having an archaeologist on site during excavation work.
- Careful recording when taking structures apart, so that they can be rebuilt in the same way.
- Providing extra protection (such as more fencing or physical protection with bubble wrap or impact-resistant sheeting) to existing buildings to avoid accidental damage from vehicles and plant.
- The correct processes to follow if you uncover previously unknown architectural details, decoration or items within buildings (such as wall paintings).
- The legal reporting process you must follow if you uncover human remains or treasure.

CONTENTS

Demolition

What your site and employer should do for you	178
What you should do for your site and employer	178
Introduction	179
Planning the work	179
Safe systems of working	179
Health hazards	180
Personal protective equipment and respiratory protective equipment	181
Plant and equipment	182
Lifting operations	183
Confined spaces	183
LPG, other gases and substances	184

DEMOLITION

What your site and employer should do for you

1. Give you a written demolition plan identifying the correct sequence and methods of demolition to prevent accidental collapse.
2. Give you suitable instruction, resources and supervision.
3. Provide you with a safe system of work, with sufficient risk assessments and method statements indicating the safe procedures to be followed.
4. Provide suitable and sufficient welfare facilities, first-aid resources, personal protective equipment (PPE), respiratory protective equipment (RPE), plant and equipment.
5. At no time place you in danger, expose you to uncontrolled hazards or expect you to take unnecessary risks.
6. Give you training so that you are competent and can work safely and in safety.

What you should do for your site and employer

1. Follow the agreed safe systems of work, keep to the requirements of the risk assessments and method statements, and work safely at all times. Where you have concerns about your safety, stop work and tell your supervisor.
2. Not take risks with your own or anyone else's health or safety.
3. Report anything that you feel is unsafe or a threat to health.
4. Keep to any permit systems, site rules and procedures that are in operation and do not operate plant unless you are qualified and authorised to do so.
5. Attend inductions, briefings, training and health surveillance as requested.
6. Look after and wear the PPE and RPE required for the job.

DEMOLITION

Introduction

Demolition is a highly specialised and potentially dangerous activity. Health and safety law requires that it is carried out by trained and competent demolition contractors.

Demolition work is any demolition, dismantling or structural alteration of any size, ranging from the explosive demolition of a whole tower block down to the removal of a single wall in a domestic house. It is not always on a large scale.

- Before starting any demolition project, the risk assessment and method statements must be reviewed to identify the intended way of controlling the hazards.
- A competent person must be appointed to supervise the work before it starts.

Planning the work

The demolition contractor must make sure that the work is planned and carried out in such a way that it avoids danger or, where this is not possible, reduces the danger as far as is reasonably practicable. This will include the following requirements.

- Identifying the location of any asbestos, before work starts, through a refurbishment or demolition survey carried out by a trained and competent asbestos surveyor.
- Identifying, locating and isolating all services (gas, water and electricity).
- Identifying hazardous materials or substances and planning for their removal and disposal.
- Confirming the arrangements for health and safety in writing before the work starts (construction phase plan).
- Setting out the sequence of operations in a written demolition plan and making sure that it is read out during site induction and understood by all involved.
- All people involved in planning, managing and carrying out demolition work must be competent (have the right blend of skills, knowledge, training and experience).

Safe systems of working

Given the hazardous nature of demolition work, it is essential that safe systems of work are developed and followed.

- Make sure that the site is secure to prevent unauthorised access, particularly by children, outside working hours.
- Make sure that there is enough clear signage (for example, indicating the boundary of exclusion zones).
- Make sure all known and unrequired services are isolated, disconnected and made dead. If underground services that were not identified earlier are discovered, work must stop until a safe system of work has been agreed.
- Speak to the appropriate service company if work has to be carried out near to underground services (such as gas, water and electricity) and/or near overhead cables.

DEMOLITION

- Work at height must be avoided where possible. If working at height is unavoidable, it should be carefully planned and carried out to prevent the fall of people and control the fall of materials.
- If demolishing internal brick walls, you must work across in even courses, from the ceiling down.
- A gas-free certificate and/or a hot-work permit must be obtained before carrying out the demolition cutting of any fuel, slurry or waste tank.
- Gas-free certificates and hot-work permits are normally only valid for a single eight-hour shift after being issued.
- Do not allow combustible waste materials (demolition soft strip materials, such as wood, paper and plastics, arising from the works) to build up.

Health hazards

Due to the nature of the working environment, demolition can present some potentially serious, long-term threats to the health of the people involved. Health issues are often overlooked because it can take many years before the harmful effects on the individual and the symptoms of disease become obvious. Workers moving between different projects and employers is another reason why long-term health can be difficult to monitor.

To help you and your co-workers stay healthy and safe on the job, make sure you understand the following points.

- Lead-based paint might be present in older buildings or under asbestos clad or coated steelwork.
- If you are involved in the hot cutting of coated steel you must be made aware of the potential for exposure to lead paints that can lead to harmful levels of lead in your blood.
- If you have come into contact with lead you must wash your hands and face before eating, drinking or smoking.
- Where asbestos is suspected of being present, a demolition survey must first be carried out so that the asbestos may be safely removed or otherwise encapsulated and protected, before any demolition work begins.
- If, during demolition, you suspect that some asbestos-containing material or other hazardous substance (for example, old lead-painted steelwork) remains in the structure, you must stop work immediately and tell a supervisor.
- Control of substances hazardous to health (COSHH) assessments must cover any other harmful substances likely to be released. As well as chemicals or fuels, it includes dusts, fumes, bacteria and so on.
- Demolition is likely to release airborne dusts and fumes (such as silica, dust and lead fumes). *(The need for respiratory protective equipment (RPE) is covered later in this section.)*
- Skin diseases (for example, dermatitis) must be prevented by making sure that there is no skin contact with harmful substances.

DEMOLITION

- Exposure to vibration in your hands and arms can lead to hand-arm vibration syndrome (HAVS).
- The safe system of work must prevent exposure to hand-arm vibration, which can be caused by using equipment (for example, breakers or drills).

Examples of tools that could cause hand-arm vibration if control measures are not implemented

Personal protective equipment and respiratory protective equipment

Wearing the correct personal protective equipment (PPE) and respiratory protective equipment (RPE) helps to prevent exposure to harmful substances. The wrong type of PPE or RPE is likely to be useless. A competent person must carry out a risk assessment to make sure the correct grade and type of RPE is selected and worn. Your supervisor must make sure that the correct PPE and RPE identified in the risk assessment are being worn and fit correctly. You must also be given a face-fit test to make sure the RPE is fitted correctly to your face. If you have a beard, you may need to be given a full-face mask. When not in use, RPE should be stored safely. Disposable RPE will lose effectiveness over time and should be replaced before becoming heavily contaminated or after each shift. Reusable RPE must have filters replaced at regular intervals, in line with the manufacturer's instructions.

- Where disposable half-mask respirators are identified in the risk assessment, they should have a minimum FFP3 rating.
- If you are working in a toxic or low oxygen atmosphere, compressed airline breathing apparatus or self-contained breathing apparatus is needed.
- An air-fed respirator or helmet will be needed if you are cutting coated steel work.

Respirators are not suitable for low oxygen environments. You must be issued with breathing apparatus.

Anyone using RPE must be face-fit tested and instructed on its maintenance and use.

DEMOLITION

Plant and equipment

Anyone who has to operate any item of plant or equipment (for example, a mobile elevating work platform (MEWP)) must be trained, competent and authorised.

- Information on the daily checks for mobile plant can be found:
  - in the manufacturer's handbook
  - in information provided by the supplier of the machine
  - on information stickers attached to the machine (sometimes referred to as decals).
- Full daily checks must be carried out by the operator as instructed in the manufacturer's handbook and/or instruction manual that was supplied with the machine.
- All plant movements and traffic routes must be planned and co-ordinated to make sure of the following.
  - There is safe access to and egress from the site (entrances and exits).
  - Pedestrians and operatives carrying out work are safe.
  - There is no conflict between different types of vehicle.
  - The movement of materials is safe.
- Any item of plant that is defective must be isolated so that no-one else can try to use it (this is known as *quarantine*), and its condition must be reported.
- Machines that operate in areas where there may be falling materials must be fitted with a falling object protective structure (FOPS).
- Roll-over protective structures (ROPS) must be fitted to plant to protect the driver where there is a danger of the plant rolling over.
- The head and tail lights and obstruction lamps of any machine must be switched on, especially when the plant is operating in conditions of poor visibility.
- Operators must never try to move a machine if they do not have adequate visibility from the driving position (such as losing sight of the vehicle marshal).
- All plant must be inspected, with the details recorded, at least every week.
- Operators must face the machine and maintain three points of contact when climbing down from it.
- Passengers must only be carried on construction plant if a purpose-made passenger seat and passenger restraint (seat belt) are fitted.
- Unattended mobile plant must be left in a safe place, with the keys removed and the doors locked.

Machine with cab fitted with falling object protective structure (FOPS)

DEMOLITION

Lifting operations

All lifting operations and use of equipment must be thoroughly planned, especially when carrying out demolition works, when there is an increased risk of falling (unsupported) materials. Poorly thought out lifting operations and the use of defective or inappropriate equipment have been the cause of many accidents.

- A lift plan (method statement) must be drawn up for all lifting operations.
- All lifting operations must be carried out within the rated capacity (safe working load (SWL)) of any item of lifting equipment or lifting accessory used.
- The rated capacity (SWL) must be marked on each item of lifting equipment and each lifting accessory (such as chains, shackles, slings and strops).
- Lifting equipment used for lifting people, and all lifting accessories, must be thoroughly examined every six months. All other lifting equipment should be thoroughly examined every 12 months.
- If an item of lifting equipment or a lifting accessory is found to be defective it must be isolated (quarantined) so that no-one can use it.
- Anyone not involved in the lifting operation should be kept out of the area.

Confined spaces

Working in confined spaces, such as oil- or chemical-storage tanks, is a potentially hazardous activity. Many people have died through lack of planning, training or access to the correct equipment. People have also died when trying to rescue others because of a lack of emergency planning, training or suitable equipment.

 No-one should be involved in confined space work unless they have been fully trained, are assessed as medically fit for the task, and are properly equipped and authorised.

- Before entering a confined space a suitable and sufficient risk assessment and safe system of work, and emergency rescue plan, must be drawn up. Everyone who is going to be involved in the operation must be fully informed of, and understand, the safe system of work, including emergency rescue procedures.
- During confined space working, it is essential that the people doing the work strictly follow the safe system of work.
- Depending on the situation (the reason for the entry), the task to be done and the location of the confined space, the safe system of work may require the issue of a gas-free certificate, permit to enter or to work, and/or other permits.

Carbon dioxide extinguishers must never be used in a confined space as the gas replaces the oxygen in the air and creates a higher risk of asphyxiation.

DEMOLITION

LPG, other gases and substances

The nature of demolition work means it is likely that highly flammable and explosive substances will be stored and used on site at some time.

- Liquefied petroleum gas (LPG) cylinders that are used for heating or cooking in a site cabin must be stored outside the cabin.
- Gases must be stored in a secured container, a safe distance away from other gases (for example, oxygen cylinders must be stored more than 3 m away from LPG cylinders).
- The correct type of fire extinguisher must be available to aid escape where petrol or diesel are being stored. Water extinguishers must **not** be provided.
- Flashback arrestors must be fitted between the pipes and gauges when using oxypropane cutting equipment.
- Cans or drums of fluids must be stored in bunded areas (a method of containing any substance that leaks from its original container) to prevent any leaks from spreading.
- If unlabelled drums or containers are discovered, work must stop until they have been safely dealt with.

LPG store that can be secured and is in a well-ventilated area, on a flat surface

CONTENTS
Highway works

What your site and employer should do for you	186
What you should do for your site and employer	186
Signing, lighting and guarding	187
Safe works – basic principles	187
Site layout	188
Traffic management and control systems	189
Mobile plant	191

HIGHWAY WORKS

What your site and employer should do for you

1. Develop safe systems of work that offer you maximum protection from road users and hazards.
2. Train you in all your work activities and give you specific training if you work on any live highway.
3. Train you, or make sure you are competent, on each specific item of equipment or mobile plant that you need to operate.
4. Provide, maintain and inspect the plant and equipment.
5. Give you the correct personal protective equipment (PPE) for tasks and high-visibility clothing for the road type.

What you should do for your site and employer

1. Follow the safe system of work and the rules for site and highway traffic.
2. Position signing, lighting and guarding in the right sequence and in the correct place.
3. Do not work in the safety zone and do not use hand signals to control traffic.
4. Wear your task-specific PPE (for example, high-visibility clothing and protective eye wear) at all times.
5. Report any defects and complete any required daily and weekly inspections of signing, lighting, guarding and equipment.

HIGHWAY WORKS

Signing, lighting and guarding

The working environment on the highway will involve you working with, or alongside, pedestrians and moving traffic and may lead to problems arising from confusion, conflict and delays.

To reduce these problems, a clear and concise **signing, lighting and guarding** procedure must be put in place.

Temporary traffic management (TTM) warns, informs and directs the pedestrian and the road user through and around the site by the means of signs, cones and barriers. Most of the common situations are described in the Code of Practice *(Safety at Street Works and Road Works)* – the *Red Book*. Further advice can be found in Chapter 8 of the *Traffic signs manual*.

Safe works – basic principles

- To comply with health and safety legislation, a safe system of work for signing, lighting and guarding will need to be in place. The safe system of work must be based on a suitable and sufficient risk assessment.

- It is management's responsibility to provide equipment in good condition – it is your responsibility to use it in the correct way.

- It is your responsibility to sign, guard, light and maintain your works safely.

- You will need to wear high-visibility clothing at all times when you are on site.

- Drivers must be able to see the advance warning signs. Where visibility is poor, or there are obstructions, extra signs should be provided. Signs should be set out for traffic approaching from all possible directions.

- You may have to duplicate warning signs on both sides of the road (for example, where signs on the left-hand side are hidden by heavy traffic).

- You must include the works area, working space and safety zone in the area to be marked off with cones (and lamps, if necessary). Never use a safety zone as a work or storage area.

- If there are temporary footways in the carriageway or obstructions (such as stored materials or plant not already within the working space), sign and guard them separately to the same standard.

- Signs, lights and guarding equipment should be secured by bags of granular material placed at low level, to avoid them being moved by the wind or passing traffic.

- Check regularly, at least once every day, that signs and cones have not been moved or become damaged or dirty.

- In many cases, traffic control will be necessary (for example, temporary traffic lights or stop/go boards).

HIGHWAY WORKS

- Traffic conditions may change from those expected and alterations may be needed. If in doubt speak to your supervisor.
- On finishing the work make sure that all plant, equipment and materials are removed quickly from the site. It is a legal requirement that all signs, lighting and guarding equipment must be removed immediately when they are no longer needed.

Works vehicles

All works vehicles should be a bright colour, so they can be easily seen, and must have an amber warning beacon. Any vehicle entering a site must switch on the amber beacon. This reduces the risk of being followed into the site by private vehicles; if this does happen, you will need to help the driver of the private vehicle to leave the site via the nearest safe designated exit.

Site layout

The site layout consists of the following.

- Advance signing – length depends on the speed and type of road.
- Works area.
- Working space.
- Safety zone, including the following.
    - Lead-in taper.
    - Longways clearance.
    - Sideways clearance.
    - Exit taper.

Setting out signs

The safety zone is provided to protect you from the traffic and to protect the traffic from you. You may only enter the safety zone to maintain cones and other road signs. Materials and equipment must not be placed in it. The sideways clearance is the space between the working space and the moving traffic and varies with the speed limit. If pedestrians are diverted into the carriageway, you must provide a safety zone between the outer pedestrian barrier and the traffic. If the carriageway width does not permit full sideways clearance you must speak to your supervisor. It may be necessary to divert traffic or reduce speeds to below 10 mph.

HIGHWAY WORKS

 Don't work in the safety zone – you may lose more than your hat.

Setting out the site

You are at greatest risk when setting out the site, so you must make sure the following precautions are in place.

- Switch on your flashing beacon(s).
- Stop the vehicle in a safe place.
- Put on your high-visibility clothing.
- Get out of the vehicle on the safest side (often the passenger side) where possible.
- Observe traffic movement at all times.
- Position signs in the correct sequence, at the correct distance and where they can be seen clearly – they must not cause a hazard to pedestrians.
- Secure the signs with sandbags or weights.
- Check that all of the signs are correct before starting work.

Traffic management and control systems

Pedestrian movement

Footway working may mean the re-routing of pedestrians. You may need to provide a temporary footway, of a minimum width as specified in the Code of Practice, using barriers (with tapping rails for the blind or visually impaired), ramps and information signs.

Motorways and high speed dual carriageways (50 mph and above)

Extra precautions are necessary when working on motorways and high speed dual carriageways, including the following.

Footpath diversion showing the safe route for pedestrians

- All traffic management must be undertaken by a registered traffic management contractor.
- Advance warning signs need to be duplicated on the central reservation.
- If you are entering a site on a motorway, you must switch on your flashing amber beacon and the correct indicator about 200 m before the access point, to give following traffic enough advance warning.
- Long-sleeved high-visibility clothing must be worn.

HIGHWAY WORKS

Short stop/mobile working operations
These include continuous mobile operations (such as hedge trimming) as well as those that involve movement with periodic stops (for example, gully emptying) and short duration works (for example, pothole filling). This work must only be carried out where there is good visibility and during periods of low risk (such as light traffic flow).

The following basic requirements for the works vehicle are the minimum traffic management needed for short stop or mobile operations.

- It must be brightly coloured so that it can be easily seen.
- It must have at least one roof-mounted amber flashing beacon operating.
- A directional, keep right/left arrow sign must be displayed on or at the rear of the vehicle, showing drivers approaching on the same side of the carriageway which side to pass. This directional sign must be covered or removed when travelling to and from the site.

Advance warning signs are necessary when there is not enough space for two-way traffic to pass the works vehicle or where it cannot be seen clearly. The signs should be placed up to one mile from the works vehicle.

Under no circumstances should you use hand signals to control traffic. Only the police are legally allowed to do this.

 The *Red book* describes the various traffic and pedestrian systems, where and when they may be installed, and who should be spoken to in all cases.

Give and take (where traffic congestion is unlikely)
These systems can be considered under the following circumstances.

- Speed limit 30 mph or less.
- Coned off area 50 m or less.
- Drivers approaching the works can see at least 50 m beyond the end of the works.
- The amount of traffic movement, including heavy goods vehicles, is low.

Priority signs
These systems can be considered under the following circumstances.

- Coned off area 80 m or less.
- Drivers approaching the works can see at least 60 m beyond the end of the coned area for a 30 mph speed limit. (Other distances are given for different speed limits.)
- Two-way traffic flow is light.

Give and take systems and priority sign systems are not suitable for use at night or when visibility is poor.

HIGHWAY WORKS

Stop/go boards
These systems can be considered under the following circumstances.

- Works length can be up to 500 m depending on two-way traffic flow levels.
- Normally use stop/go boards at each end of the works.
- Where visibility is poor, a communication system must be used.
- Allow enough time for traffic to clear with both boards showing 'Stop'.
- Speak to your supervisor if the works are near a railway level crossing or a road junction.

Portable traffic signals (temporary traffic lights)
These systems can be considered under the following circumstances.

- Works length can normally be up to 300 m.
- Signals must be put up and removed in an organised way and specific sequence.
- Allow more time for slow-moving traffic, cyclists and horse riders by increasing the all-red timings.
- Most sites will only need one set of signal heads. Where visibility is poor, a double-headed system should be used.
- Where signal cables cross the carriageway, cable protectors must be used and signs indicating 'Ramp' or 'Ramp ahead' must be used.
- If the detector systems become faulty, you must operate the signals on fixed time or manually, and contact the service company or your supervisor.

Mobile plant

The movement of mobile-operated plant and vehicles causes many accidents and serious injuries on sites every year. The accidents involve not only the operator but also people close by or walking past.

 Mobile plant **means all mobile plant and site vehicles that can move either under their own power or by being towed.**
Operator **means anyone driving or operating mobile plant.**
Pedestrians **means anyone on foot.**

HIGHWAY WORKS

Maintaining mobile plant

If you are an operator you are responsible for daily and weekly maintenance to make sure your plant and/or equipment is in a safe condition, giving particular attention to the following.

- Checking levels of fuel, oil, water and brake fluid.

- Ensuring the horn, reversing alarms, flashing beacon, headlights, indicators and any pedestrian proximity sensors and/or cameras all work.

- Making sure that brakes, including hand and parking brakes, are operating efficiently. Where air brakes are fitted the air storage tanks must be drained at the start of every shift.

- Making sure windscreen wipers and washers operate efficiently, including keeping the washer bottle topped up.

- Keeping windows and mirrors clean, and correctly adjusted for good visibility. This is particularly important when manoeuvring.

- Keeping the cab clean, tidy and clear of any loose articles that may obstruct the operation of foot pedals and controls.

- Checking that your clothing, especially if you are wearing a jacket, does not snag on controls, isolators or safety switches.

Pre-use checks will help to make sure your vehicle is safe

Diesel must never be used to clean mobile plant or prevent bitumen or asphalt sticking to buckets or load beds, as this will make things slippery and present a serious risk of injury from slips and falls. Use suitable access equipment to prevent serious injury from falls when hosing down plant, especially for high level or load beds of lorries and gritters. This type of work is classed as working at height, as there is a risk of injury from a fall.

Mobile plant in an unsafe condition must not be used.

Operating mobile plant

If you operate plant you must be trained and competent for each type of plant you are authorised to use. You should have a suitable driving licence and/or approved plant operator's certificate or card. Before operating any plant for the first time, you should read and understand the manufacturer's operating instructions and be familiar with the controls and their function.

- You must not work too many hours (the Drivers' Hours Regulations and Working Time Regulations limit the number of hours that may be worked in any day or week).

HIGHWAY WORKS

- Operator and plant record books must be completed, where necessary.
- You must be aware of the gross vehicle weight, the maximum axle weights and overall dimensions of the plant.
- When refuelling any plant, remember that this should be carried out in a designated area or with suitable protective measures to avoid pollution being created. In addition, there must be no smoking or naked lights and you should switch off the ignition.
- You are responsible for the safe operation and condition of your mobile plant at all times.
- You must comply with road traffic legislation and site rules where applicable.
- The use of an amber beacon does not exempt you from compliance with the Highway Code.
- You must stop operations immediately and report defects that present a serious risk to the safe operation of plant (for example, faulty controls).
- Where fitted, seat belts must be worn (they could save your life).
- Consider the need for a vehicle marshal when manoeuvring, and use mirrors and CCTV, where fitted.
- A vehicle marshal must be used when reversing in areas where there may be pedestrians.
- Do not operate mobile plant too close to any excavation, no matter how shallow.
- Stop blocks are the preferred method of preventing mobile plant getting too near to an excavation when tipping.
- Obey all speed limits, height restrictions, direction and warning signage.
- Be alert to the dangers of colliding with or clipping overhead cables, scaffolding, temporary works, mobile towers, mobile elevating work platforms (MEWPs) and ladders.

Parking

- Where possible, park on level ground in a designated area, clear of pedestrians. The handbrake/parking brake should be on, the engine turned off and the key removed.
- If you cannot park on level ground you may need to chock the wheels or otherwise prevent unintended movement of the plant.
- All hydraulic equipment (such as buckets, forks and back-actors) should be lowered to the rest position.

HIGHWAY WORKS

Access
- Only authorised people should be allowed on mobile plant.
- Passengers may only be carried on mobile plant that is equipped with sufficient, suitable seating for them.
- Wait until the mobile plant has come to a complete stop before getting on and off. Never jump down.
- Maintain three points of contact and always face the access ladders or hand and foot holds when getting into or out of the machine.
- Never, at any time, work under an unpropped mobile plant body.

 Maintain three points of contact – don't jump down.

The following operations may need detailed consideration and risk assessment to work out safe access.
- Maintenance.
- Working at height at the top of the mobile plant, including the sheeting of loads.
- Anywhere else where falls may be likely.

Loads
Loading and the load itself are the vehicle marshal's and operator's responsibility.
- Plant must not be overloaded and loads must be spread evenly and be secured.
- Operators of mobile plant should not stay in the seat or cab while it is being mechanically loaded.
- Operators must check that sideboards, curtains, sheeting and tailboards are fastened and secure before moving off.
- Operators must check the load for security. Any projections must be properly marked and clearly visible.
- Several incidents and deaths have occurred during the unloading of vehicles. Care should be taken when removing lashings as the load may have shifted during transit, causing the load to collapse on being released.
- Loading and unloading of tippers must be attended by a competent vehicle marshal. Rear end tippers are at risk of overturning when tipping on uneven or unmade ground.
- People should stay well clear of any tipping or loading operations.
- Tippers and dumpers must not travel with the bucket in the raised position after the load has been discharged.

HIGHWAY WORKS

Trailers

Before towing any trailer or plant, you must make sure you have the correct class of driving licence. It is essential that the driver is licensed for the specific combination (towing vehicle plus trailer and load).

- The towing vehicle and trailing vehicle must be compatible and properly secured with the correct towing pin. Non-standard couplings are not acceptable.
- When being towed on a highway the trailer must be fitted with registration number plates, indicators, brake and rear tail lights.
- Where fitted, trailer parking brakes must be applied before disconnecting the trailer from the towing vehicle. On slopes, trailer wheels should also be chocked.
- Trailers fitted with independent operating brakes should be connected to the towing vehicle by a cable, which will activate the trailer's brakes if the tow hitch fails. Otherwise the trailer must be fitted with a safety chain connected to the vehicle.

Loads should not be secured to sheeting (rope hooks) as they have no rated capacity (SWL). Loads must only be secured to designated load-securing points and rings, using appropriate chains and strops.

Mobile plant used for lifting

- All lifting operations must be properly planned by a competent (appointed) person.
- Only authorised and trained operators should operate lifting equipment.
- All lifting equipment and lifting accessories should be subject to a daily and pre-use visual inspection.
- Care must be taken when operating extending booms or other lifting equipment near to overhead power lines. The safest method is installing overhead cable goalposts, limiting the boom extension or height of the mast so it does not go into the overhead exclusion zone.
- Equipment for lifting people (such as a MEWP) and lifting accessories must be thoroughly examined every six months.
- Other lifting equipment must be thoroughly examined every 12 months.

All lifting operations must be properly planned

The rated capacity (SWL) of any lifting equipment must never be exceeded.

HIGHWAY WORKS

CONTENTS
Specialist work at height

What your site and employer should do for you	198
What you should do for your site and employer	198
Introduction	199
Hierarchy for working at height	199
Safe working at height	200
Roof work	201
Means of access	202
Fall arrest (protecting you if you do fall)	207
Preventing falling objects	209

SPECIALIST WORK AT HEIGHT

What your site and employer should do for you
1. Make sure work at height is properly planned so that your place of work and the access route to and from it are safe to occupy.
2. Provide, inspect and maintain the most suitable work at height equipment.
3. Provide adequate safety equipment to prevent falls and, if necessary, to arrest any falls.
4. Give you a written safe system of work and confirm your understanding of it, so that you can work at height safely.
5. Make sure you are trained and regularly updated in methods of working safely at height.

What you should do for your site and employer
1. Fully understand and follow the agreed safe system of work (including using any equipment provided to prevent or arrest falls).
2. Do not use powered access (or any other) equipment unless you have been trained and authorised.
3. Not interfere with anything provided for safety.
4. Not take risks or short cuts.
5. Report any falls which occur, even if they cause no injury.
6. Report any unsafe aspect of your work or work conditions that you feel may be unsafe.

SPECIALIST WORK AT HEIGHT

Introduction

- Every year many deaths and serious injuries occur because of falls from height.

- All too often the same types of accident reoccur (such as falls through fragile roof materials and falls from ladders).

- In many cases good planning, taking basic safety precautions and/or providing training could have prevented suffering.

 You should also read Chapter D17 Working at height, which explains what employers and, in some cases, you must do to comply with the law and keep yourself and others safe whilst working at height.

Hierarchy for working at height

Any person planning work at height should always follow a hierarchy of control and consider options at the top of the hierarchy before moving down.

Step 1. Avoid working at height
e.g. assemble on the ground and lift into position using a crane or by fixing guard-rails to structural steelwork on the ground before lifting and fixing at height

⬇

Step 2. Prevent falls from occurring
Use an existing safe place of work
e.g. parapet walls, defined access points, a flat roof with existing edge protection

⬇

Step 3. Prevent falls by providing *collective* protection
e.g. scaffolding, edge protection, handrails, podium steps, mobile towers, MEWPs

⬇

Step 4. Prevent falls by providing *personal* protection
e.g. using a work restraint (travel restriction) system that prevents a worker getting into a fall position

⬇

Step 5. Minimise the distance and/or consequences of a fall using *collective* protection
e.g. safety netting, airbags or soft-landing systems

⬇

Step 6. Minimise the distance and/or consequences of a fall using *personal* protection (The last resort)
e.g. industrial rope access (working on a building façade) or a fall-arrest system (harness and fall-arrest lanyard), using a suitable, high level anchor point

SPECIALIST WORK AT HEIGHT

Safe working at height

Safe working at height depends upon the following requirements being met.

- The work is properly planned with a safe system of work in place, which follows the hierarchy of control for working at height.
- The people who will be working at height are adequately trained, competent and authorised.
- The job is carried out in line with the safe system of work.
- Adequate supervision is provided.
- The tools and equipment to be used are properly maintained and inspected.

Before you start working at height (for example, on a roof) there are many things that your employer, or you, should do to make sure that the job is carried out safely. Some examples are shown below.

- Assess the risks arising from the work and take measures which will:
  - prevent people or materials falling from height
  - protect people and property from serious injuries or damage if someone or something does fall.
- Draw up a safe system of work that explains exactly how the job should be carried out and make sure that the people doing the job understand it. The safe system of work should include details of the following.
  - How falls will be prevented or arrested.
  - The equipment to be used and the competence needs of operators.
  - Who will supervise the job.
  - The sequence of operations that must be carried out.

If you are not sure that what you have been asked to do is safe, or if you are unhappy about any other aspect of working at height, you should talk about it with your supervisor.

Generally, the risks will depend on the following.

- The height and nature of the place of work (for example, a scaffold platform or leading edge of a new roof).
- The method of access chosen to get the people doing the job to the place of work.
- Details of the safety precautions and rescue plan in case someone does fall.
- The number of people who will be working at height.

Working at height in a safe environment

SPECIALIST WORK AT HEIGHT

- The length of time that people will be working at height.
- How experienced the people are in doing the job they need to do.
- The nature of any materials that will have to be hoisted and/or stored at height.
- The method of hoisting materials.
- The weather.
- Whether or not people can be kept out of the area by putting exclusion zones below where the work is taking place.

 Falls from height are still the biggest killer, but many could easily have been prevented.

Roof work

Much work that is carried out at height involves working on roofs, which can be hazardous.

- Roof work is particularly dangerous because it is not always clear at the start whether or not the roof cladding is a fragile material. A safe system of work must be in place for all work at height.
- Fragile roof lights can be dangerous if not protected with securely fixed covers that can withstand heavy loads imposed on them.
- Remember that surfaces can become slippery after rain, frost, snow and where moss or lichen is present.
- High-level features (such as overhead cables) present further dangers, such as electrocution.
- If you are working near power cables check with your supervisor or the person in charge of the site that the cable is isolated and it is safe to be on the roof.
- Specialist access equipment is likely to be needed for carrying out work above fragile roofs (for example, the inspection of pipework).
- Materials that have to be stored on roofs must be stored in a safe way so that:
  - they do not pose a hazard to anyone working on the roof
  - the pitch of the roof is taken into account when deciding what can be stored safely
  - a safe method of getting the materials up to the roof is used (for example, inclined hoist, scaffold-hoist, safety pulley or gin-wheel)
  - excessive 'point loading' is avoided during loading-up
  - the materials cannot fall or be blown off the roof
  - the materials can be accessed safely
  - the materials can be safely and quickly distributed around the roof to where they are needed.

SPECIALIST WORK AT HEIGHT

If you are planning to use a roof ladder, consider the following.

- Whether a roof scaffold is a better option.
- A stable working platform at eaves height will provide safe access to the roof ladder with the added feature of edge protection.
- If a roof ladder is to be used, it must only be used for tasks of a short duration.
- If using a leaning ladder for access, the transition between any two ladders must be safe.
- A leaning ladder used for access must be stable and extend 1 m above eaves height.
- The top of the roof ladder must be secured with a properly designed roof hook on the opposite slope of the roof and not rely on the ridge tile for its anchorage.

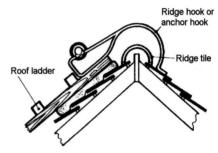

Typical roof ladder in use

Means of access

Scaffolds

A common way of preventing falls is to use a secure and stable working platform (such as a scaffold). Below are some essential considerations when using scaffolds.

- Is it a safe and suitable means of access for the job that has to be done (for example, are the lift heights satisfactory)?
- Scaffolds must only be erected, altered and dismantled by a scaffolder who is trained and competent. The scaffolder should hold a current industry-recognised Construction Industry Scaffolders Record Scheme (CISRS) card.
- Edge protection must be fitted to all working platforms.
- Each working platform must be wide enough to allow the job to be carried out safely, and allow for the passage of people and equipment as necessary.
- Every scaffold must be inspected periodically (normally every seven days) by a competent person, with the findings of the inspection entered in a register. Details should also be recorded on a plastic tag, usually fixed to the scaffold next to the access point.
- Never overload working platforms; this has been the cause of many scaffold collapses.
- Untrained, unauthorised people should **never** interfere with a scaffold in any way (for example, by removing a tie or guard-rail which is in their way).

SPECIALIST WORK AT HEIGHT

Scaffold prohibition tag   *Scaffold inspection record*

 If you notice any unauthorised modifications to a scaffold, report it to your supervisor straight away.

Mobile elevating work platforms

Common types of mobile elevating work platform (MEWP) are cherry pickers (boom type) and scissor lifts. Safety considerations when using a MEWP are shown below.

- You must not operate any type of MEWP unless you are trained, competent and authorised. Holding an International Powered Access Federation (IPAF) card is a good indication of training, and may be insisted on by many employers.

- Users of cherry pickers must normally wear a safety harness and restraint (fixed-length) lanyard secured to the designated anchorage point, unless a documented risk assessment dictates otherwise.

- The MEWP chosen for the task must allow safe access to the place of work. Do not stand on the guard-rails or put a stepladder or hop-up on the working platform. If the MEWP does not reach, use a larger machine.

- Operators must know how to carry out daily and weekly checks.

- The person planning the task must make sure that a survey is carried out to identify any overhead hazards (such as power lines) or underground voids (such as drains or cellars) which could collapse under the load.

- Only use a MEWP in areas identified in your lift plan, as ground conditions may not have been assessed in other areas.

SPECIALIST WORK AT HEIGHT

- Before elevating the machine, operators must identify any projections or other features on the structure which could trap them between the guard-rails and the structure.
- Wind speed forecasts should be obtained before work commences.
- The Beaufort Scale gives managers and workers indicators of prevailing wind conditions. The scale categorises the wind strength between Force 0 and Force 9. The higher the number, the higher the wind speed (for example, Force 2 is a light breeze and Force 7 is a near gale).
- Operators must be aware of the wind-loading on a raised MEWP and be prepared to stop work and lower the machine if the wind speed is judged to be too high.
- There must be a rescue plan in place in case the operator is stuck or injured and cannot lower the working platform.
- The ground-level controls must only be used in an emergency (for example, if the operator becomes ill or is trapped).
- A MEWP must not be used as a substitute for the stairs in the structure being worked on.
- If working near water, consider whether the operator needs to wear a harness. Evaluations should be carried out to decide if the greater danger would be from drowning, in the event that the operator becomes trapped underwater by the harness.

Auxiliary controls decal

Note: *if there is an emergency, a responsible person on the ground should know where to find and use the lowering controls (sometimes shown by an emergency descent symbol).*

 Never carry more than the maximum safe load in a MEWP. The maximum load in kilogrammes or number of people will be shown in your lift plan and displayed on the machine.

Mobile access towers

- Mobile access towers can be static or mobile, depending if wheels are fitted.
- They can be built from tube and fitting scaffold components or, more commonly, prefabricated alloy frames which slot together.

Tube and fitting towers must only be erected, altered or dismantled by a trained and competent scaffolder (for example, someone who holds a CISRS card).

Alloy towers must only be built, altered or dismantled, in line with the manufacturer's instructions, by someone who has been trained and is competent on that type of tower (for example, someone who holds a Prefabricated Access Suppliers and Manufacturers Association (PASMA) card).

SPECIALIST WORK AT HEIGHT

Listed below are some safety considerations.

- The ground or floor surface must be level and sufficiently firm to take the loading of the base plates (static) or wheels (mobile).
- The guard-rails and toe-boards must be fitted before the tower is used.
- Some types of alloy towers have guard-rails that can be positioned before the working platform is accessed.
- The brakes must be on at all times when a mobile tower is in use.
- Once the platform of a mobile tower has been occupied the trapdoor must be closed immediately to prevent anyone or anything falling through it.
- Provision must be made for the safe hoisting of tools and materials up onto the platform.
- Every tower must be inspected periodically by a trained, competent and authorised person, including an inspection after any event that is likely to have made it unsafe to use (such as impact by a vehicle or being subjected to severe wind or weather).
- The working platform must only be accessed by using the built-in ladder (**never** climb up the outside of a tower or use a free-standing ladder).
- A mobile tower must not be moved whilst any person(s) or equipment are on the platform.

Ladders and stepladders

More stable items of access equipment are now available for tasks that, at one time, would have been carried out using a ladder or stepladder. Ladders and stepladders should only be used when carrying out light work of short duration where the risk of a fall is low. A risk assessment will show whether a ladder or stepladder is suitable for any particular job. The risk assessment **must** also state that other options have been considered for access or as a working platform, to confirm that a ladder or stepladder is the most suitable option.

If the risk assessment shows that a ladder or stepladder is the most appropriate equipment, then consideration must be given to the following areas.

- Any ladder used on construction sites must have been manufactured for industrial use and will be labelled 'Class 1' Industrial or 'EN 131' Trade and Industrial – heavy duty and industrial use (for professional users).
- Possible defects, including splits in the material from which the equipment is manufactured, missing or distorted rungs or frayed, broken or missing tie-cords.
- Where a ladder is used to access a high-level platform, there must be a landing at least every 9 m that the ladder rises vertically.
- Make sure that an access ladder extends at least 1 m above the stepping-off point if there is not another handhold.
- Various fittings (such as those below) are available that allow ladders to be used safely.
    - Stand-off frames that avoid the need to rest ladders against fragile or flexible materials (such as plastic guttering).

SPECIALIST WORK AT HEIGHT

- Anti-slip mats and other anti-slip devices upon which the stiles are positioned.
- Adjustable stabilisers to prevent sideways slipping.
- Adjustable extensions to allow use on sloping ground.
- Ridge-hooks that allow an ordinary ladder to be converted to a roof ladder.

- Ladders are made from various materials.
  - Generally alloy ladders are lighter than wooden ones but will conduct electricity and so cannot be used close to live overhead cables and electrical equipment.
  - Fibreglass ladders are non-conducting and so are safer to use near electrical supplies or electrical equipment.

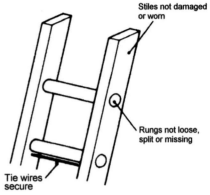

Some considerations for ladder inspections

 All types of ladder will conduct electricity if they are wet.

- If you are planning to use an extending ladder consider the following.
  - Keep the minimum overlap between sections, as given in the manufacturer's instructions.
  - The latching mechanism (and hoisting rope/fittings if appropriate) must be part of the pre-use inspection.
  - The length/weight of the ladder and how many people will be raising it will decide whether it is safer to extend the ladder before raising it.
  - Look for overhead obstructions **before** raising the ladder.

Rope access

- Rope access is a specialised activity requiring particular skills and training.
- Anyone engaged to carry out rope access must be Industrial Rope Access Trade Association (IRATA) or similar industry trained.
- Generally, rope access is needed where it is impractical to provide a working platform, other fall prevention methods or fall-arrest measures.
- Tasks that rope access can be used for include the following.
  - Structural surveys.
  - Non-destructive testing.

Rope access system (Image supplied by PETZL)

SPECIALIST WORK AT HEIGHT

- Localised concrete repairs.
- Cladding/glazing panel replacement.
- Secondary fixings.
- Surface preparation and decorating.
- Pressure pointing.

Fall arrest (protecting you if you do fall)

Health and safety law recognises that there will be situations where physical barriers cannot be used, and the use of collective fall-arrest measures (such as safety nets or airbags) is not practical. In these circumstances personal fall-arrest measures (such as a safety harness and lanyard) may be used as a **last resort**.

The following items should be considered when your employer chooses fall-arrest measures.

- Collective fall-arrest systems (such as safety nets or airbags) are preferable to personal protective measures (such as a safety harness and lanyard).

- A safety harness and lanyard should only be used in situations where:
    - it is not practical to use other fall-arrest measures (such as a safety net or airbags)
    - a purpose-designed, secure anchor point is available, either integral (built into the structure) or mobile anchor
    - users have been trained in the pre-use inspection, fitting and use of the equipment
    - users are aware that any damaged equipment must not be used
    - the workplace must be designed or secured, and the work must be planned to avoid a swing fall or pendulum effect in case someone should fall. Severe injury can occur in this type of fall
    - a rescue plan is in place to reduce to a minimum the time that someone is suspended in a harness. An injured person's condition can deteriorate rapidly if they do not receive medical attention quickly. A person not injured in a fall may suffer injury from a prolonged suspension (known as *suspension syncope*, where the suspended person faints and suffers from further health complications).

- Essential features of any harness used for fall arrest are shown below.
    - A full body harness will keep someone who has fallen in an upright position until rescued.
    - Ideally it will be fitted with a feature that allows the person to raise their thighs above horizontal or to 'stand up' whilst awaiting rescue, releasing the pressure caused by the leg straps and reducing the chances of further injury.
    - It must be inspected before use, with more detailed periodic inspections at intervals set by the employer. (It is recommended that a detailed inspection is carried out at least every six months.)

SPECIALIST WORK AT HEIGHT

Safety nets

- Safety nets should be rigged immediately below where people are working to minimise the distance a person could fall.
- Safety nets must be rigged by someone trained, competent and authorised to do so.
- Nets must be periodically inspected and any damaged nets replaced.
- If any damage to a safety net is discovered, work above the net must stop, the damage must be reported, and the net replaced or repaired by a competent person.
- Materials or debris that have fallen into the net must be quickly cleared from it.
- No-one other than a trained and competent net rigger should make any adjustments with the securing cords. If you find one of the cords is in your way, report the problem.
- Where nets are rigged using a MEWP, the floor surface must be suitable and able to withstand the loading.
- If a MEWP cannot be used it may be necessary to employ competent rope-access technicians to rig the nets (for example, IRATA-trained technicians).

Safety nets installed before roof work starts

Airbags/bean bags

These systems can be an effective method of arresting falls, if they are installed correctly. They are generically known as *soft-landing systems*.

- They must be supplied and installed by competent contractors.
- They must be clipped together as necessary to provide a continuous fall protection area.
- They must fill the area over which fall arrest protection is needed.
- Airbags must not be too big; if over-size they will exert a sideways pressure on anything that is confining them and they may not inflate as designed.
- Airbag inflation pumps must be kept running whenever anyone is working above the air bags.
- Care must be taken to make sure that anyone who falls cannot bounce or slide onto another hazard (for example, onto exposed rebar or out of a first floor window).

SPECIALIST WORK AT HEIGHT

Preventing falling objects

One of the obvious dangers of working at height is the possibility of materials, equipment and hand tools being dropped onto anyone below the place of work. Work at Height Regulations ban the throwing or deliberate dropping of any item of equipment, material or substance from a place of work at height. Measures that should be considered to prevent this are listed below.

- Being aware of the presence of people, including pedestrians, below the work area.
- Fitting edge protection to the working platforms of scaffolds.
- Storing materials awaiting use in a safe way, away from the edges of roofs.
- Consider fitting a short lanyard with a wrist-loop to hand tools.
- Wearing a chin-strap with a safety helmet to prevent it falling off if you need to bend over while working.
- Rigging a safety net, overlaid with a fine mesh.
- Using a waste chute, with the bottom just above a skip, to move waste materials to a lower level. Consider shielding or covering the skip to limit the creation of dust as a result of using the waste chute.

SPECIALIST WORK AT HEIGHT

CONTENTS
Lifts and escalators

25

What your site and employer should do for you — 212

What you should do for your site and employer — 212

Overview and control measures — 213

LIFTS AND ESCALATORS

What your site and employer should do for you
1. Make sure that a competent person plans and supervises the work.
2. Provide you with safe systems of work that are without risks to your health.
3. Explain the safe systems of work to you, using risk assessments and method statements.
4. Give you adequate information, instruction and training.
5. Give you the right personal protective equipment (PPE) and provide the right tools and equipment for the job.

What you should do for your site and employer
1. Follow the agreed safe system of work.
2. Use only equipment and methods of work you have been trained in.
3. Wear the right PPE for the task.
4. Do not take risks or short cuts.
5. Stop and seek advice if anything changes or seems unsafe.

LIFTS AND ESCALATORS

Overview and control measures

By the nature of the equipment, work on lifts and escalators can be hazardous and must only be carried out by trained, competent and authorised individuals.

The Lift and Escalator Industry Association (LEIA) represents most of the lift and escalator companies in the UK. Its members are committed to the LEIA safety charter, which requires member companies to work in line with BS 7255 and BS 7801.

The actual requirements for this training and competence are detailed in Codes of Practice *Safe working on lifts* **(BS 7255) and** *Safe working on escalators and moving walks* **(BS 7801).**

Employers should make sure that all of their employees undertake the following.

- Use and verify **stop** and other devices when accessing, exiting and working on car tops and in lift and escalator pits to make sure they have total control of the equipment.
- Electrically isolate and lock off, when power is not needed and when working close to unguarded machinery.
- Protect themselves and others from falls and falling objects.

These are three essential control measures that members of the LEIA sign up to. These measures must be carried out to make a safe working procedure.

For further guidance refer to:

Lift and Escalator Industry Association
33-34 Devonshire Street
London
W1G 6PY

Tel: 020 7935 3013

www.leia.co.uk

LIFTS AND ESCALATORS

25

CONTENTS

Tunnelling

What your site and employer should do for you	216
What you should do for your site and employer	216
Introduction	217
Safe systems of work	217
Communication	218
Safe access and egress	218
Emergency actions	219
Fire and hot works	219
Electrical safety	219
Atmosphere	220
Health risks	221
Equipment and moving plant	221
Pipework, services and hoses under pressure	222

TUNNELLING

What your site and employer should do for you
1. Make sure that a competent person plans and supervises the work.
2. Provide you with safe systems of work that are without risks to your health.
3. Explain the safe systems of work to you, which will include a risk assessment and method statement.
4. Give you adequate information, instruction and training.
5. Give you the right personal protective equipment (PPE) and provide the right tools and equipment for the job.

What you should do for your site and employer
1. Follow the agreed safe system of work.
2. Use only equipment and methods of work you have been trained in.
3. Wear the right PPE for the task.
4. Do not take risks or short cuts.
5. Stop and seek advice if anything changes or seems unsafe.

TUNNELLING

Introduction

Due to the nature of tunnelling, where people are often working far underground and a long way from a place of safety in the open air, the importance of working in a way that is safe and free of risks to health cannot be stressed enough.

This chapter highlights some of the common potential hazards that could be experienced during tunnelling operations. The risk control measures and safe systems of work identified follow BS 6164, which is the Code of Practice for health and safety in tunnelling in the construction industry.

Safe systems of work

Formal safe systems of work are essential during tunnelling operations. Some examples are listed below.

- Making sure that where a shaft is being constructed, provision is made for:
  - adequate guarding
  - continuous gas monitoring
  - ventilation.
- Understanding that during shaft sinking operations the greatest danger from suspended loads comes from falling objects or failure of the load.
- Wearing a safety harness when building rings from a platform within a shaft.
- Making sure of continuous structural stability by immediately installing supports to excavated ground.
- During tunnel construction, using compressed air (where and when necessary) to prevent or reduce the ingress of water.
- Storing tunnel segments in a secure way that prevents:
  - falling or collapse
  - any other movement
  - instability.
- Making sure that personal protective equipment (PPE), safety signs and notices, and a rescue plan are provided to everyone.
- Using water-spray curtains to help reduce the movement of smoke through a tunnel.
- Protecting against the rebound or fall of materials during and after sprayed concrete operations.

A training tunnel
(© Crossrail Ltd)

TUNNELLING

- Taking great care to make sure that oxygen cylinders do not come into contact with grease, to avoid potential explosions.
- Making sure there is a safe system of work for track maintenance workers and locomotive drivers, including the provision of the following.
    - Flashing lights either side of the work area.
    - Adequate refuge points.
    - A permit to work.
    - A lookout person.
    - Everyone involved in the job wearing high-visibility clothing.

Communication

Effective communication is essential for safe tunnelling operations. For this reason, the power supply for communications equipment must be independent of the mains power supply so that it continues working if the mains power fails.

The most common form of communication between the surface and the tunnelling face is either by radio or a tannoy system.

Lone working should **not** be allowed in tunnels because if you are taken ill or have an accident you may not be able to communicate.

It is critically important that you report any defects (such as damaged ventilation ducting) to a supervisor as soon as possible.

Safe access and egress

In tunnelling operations, getting to and from a place of work safely is as important as working safely whilst there.

Some things to consider are listed below.

- Fitting secure barriers of at least 1.2 m in height around every shaft, to prevent falls.
- Every working shaft must have a minimum of two escape methods or routes from each place where people are working.
- A tally system must be in place to control entry and exit to any tunnel under construction. A tally system counts people in and people out.
- The **maximum** distances between safe refuges in a tunnel are 50 m on straight sections and 25 m on curved sections.
- Safety nets and/or sliding doors must be fitted to equipment used for lifting people (such as cages, hoists and baskets) to prevent anyone falling out or leaning out whilst they are in motion.
- Vertical ladders in tunnel shafts must have a landing (resting place) every 6 m.

TUNNELLING

- For maintenance purposes, water must not be allowed to build up above rail level in a tunnel.

- The lifting cages on equipment used for lifting people must not be overloaded. A label or notice will state the maximum number of people that can be carried.

Emergency actions

In effect, tunnels are confined spaces that need, in most cases, a self-rescue set to be made available for each worker. Where one is provided, you should be aware of the following.

- It must be immediately available at all times.

- The duration of the air supply will be significantly reduced if you exert yourself whilst wearing it (such as when running).

- At best, the air supply will only last for 20 minutes.

In an emergency, the rescue services will use the tally-board to find out who is underground.

If there is an emergency or fire, if possible raise the alarm, and if safe to do so leave the tunnel.

If the ventilation system failure alarm activates, you should immediately evacuate the workplace.

Fire and hot works

An underground fire can have devastating consequences. The close control of fire risks (such as those listed below) is therefore essential.

- Fire checks must be maintained for at least 60 minutes after any hot works have stopped.

- Smoking, or even the carrying of smoking materials, is not allowed underground.

- Smoking or other naked flames are not allowed within **10 m** of battery charging areas because of the flammable gas produced by lead-acid batteries when they are charging.

- Hot works are not allowed within **10 m** of any diesel fuelling point or lead-acid battery charging area.

Electrical safety

Emergency lighting must be installed in tunnels and must come on if the main lighting fails. The maximum distance between emergency lighting units is **50 m**.

For personal safety, apart from battery-operated tools, all powered hand tools must operate from a 110 volt supply.

Industrial electrical cables, plugs and sockets are colour coded to show what voltage they are carrying.

TUNNELLING

Yellow	110 volt.
Blue	230 volt.
Red	400 volt.

Atmosphere

In tunnelling, a safe and breathable atmosphere cannot be guaranteed unless suitable control measures are taken.

Using the right equipment is essential to avoid creating hazardous atmospheres.

- Only use intrinsically safe (spark proof) electrical and other equipment to limit the risk of ignition of any potentially explosive or flammable atmosphere that may form.
- All diesel or electrically powered plant must be fitted with a **fixed** fire-extinguishing system.
- **Do not** use petrol-powered internal combustion engines. Internal combustion engines increase the potential for a flammable atmosphere to form and an increase in carbon monoxide.

Maintaining a satisfactory level of oxygen in the air is essential. Normally, there is 21% oxygen in the air we breathe. A lack of oxygen, which causes breathlessness, occurs if the level falls below 19%.

There are potential dangers from other gases that may be found underground.

- Exposure to **hydrogen sulphide** can kill through respiratory paralysis. It smells like rotten eggs and so its presence should be immediately obvious. However, hydrogen sulphide effectively paralyses the nerves in the nose, leading people to believe that the gas has disappeared because they can no longer smell it, when, in fact, it is still present.
- Naturally occurring **methane** is highly flammable and can combine with the oxygen in the air to form an explosive mixture.
- Exposure to **carbon monoxide** prevents the intake of oxygen into the body.
- **Nitrogen oxide**, which is produced by diesel-powered equipment, can cause breathing problems.

 The use of a calibrated multi-gas monitor is a reliable way of detecting the presence of methane and carbon monoxide.

Forced ventilation is one way of limiting the potential build up of hazardous gases underground. The failure of ventilation equipment must be indicated by an audible and visual alarm and should trigger an immediate and managed evacuation of the workplace.

TUNNELLING

Health risks

- Tunnelling work will sometimes involve working in a compressed air environment. As a consequence, decompression illness, often shortened to DCI, is a potential health risk.

- Concrete and cement batching plants must be equipped with an eyewash station to provide quick treatment to anyone who has concrete splashed into their eyes.

The following health risks are associated with sprayed concrete linings.

- Cement burns.
- Hand-arm vibration syndrome.
- Inhalation of dust.
- Hearing damage as a result of exposure to noise.

The following are some of the hazards and risks associated with hand-mining.

- Hand-arm vibration syndrome (HAVS), which is caused by a combination of the level of vibration and the duration and frequency of the exposure time. The effects of HAVS include damage to the nerves and blood vessels in the hand, which cannot be cured.

- Noise.

- Inhalation of dust.

- Roof falls (major and minor collapse of materials from above).

Equipment and moving plant

The movement or operation of plant near to where people are working will create a potentially dangerous situation unless a safe distance of separation is maintained. Working in the confines of a tunnel can make this difficult to achieve. Below are some examples of requirements and good practice.

- Crush zones caused by moving plant will be indicated by:
  - flashing warning lights
  - signs or barriers.

- All conveyors must be fitted with:
  - an audible alarm to indicate that it is about to start
  - an emergency stop system (pull cord or button).

- Inclined conveyors must be fitted with an anti roll-back device to prevent the belt running backwards if power is lost.

- Making sure that segment erectors have a fail-safe device to stop the operation if the equipment suffers any malfunction (such as developing a leak or a power supply failure).

- Fitting audible and/or visible alarms to rams and erectors of tunnel boring machines or hydraulic jacking rams in pipe-jacking shafts/pits, to warn when they are moving.

TUNNELLING

- An additional control point is the most effective way of overcoming the restricted vision of tunnel boring machine (TBM) operators during the building process.
- Always using a hop-up or refuge in a tunnel when vehicles are passing.
- Always go immediately to a hop-up or refuge if a locomotive or other vehicle is approaching.
- The lights fitted to locomotives must be:
  - visible at a minimum distance of **60 m**
  - white (if fitted) to the front of the locomotive
  - red (if fitted) to the rear of the locomotive.
- Secondary couplings must be fitted to un-braked rolling stock to reduce the risk of it running away.
- In an emergency, a locomotive must be able to stop within a distance not exceeding **60 m**.
- Ideally, the movement of locomotives when entering the back of a TBM will be controlled by:
  - traffic lights
  - closed-circuit television (CCTV) in the cab.
- The traffic light system used to control plant movement underground is as follows:
  - **Red** – stop.
  - **Amber** – out bye.
  - **Green** – in bye.

A student learning how to drive a tunnelling locomotive

Pipework, services and hoses under pressure

All services running through the tunnel should be safely positioned to avoid them being damaged. The use of pressure systems is potentially hazardous unless the safe working practices listed below are adopted.

- Isolating or releasing stored energy before disconnecting or uncoupling tunnel services.
- Including isolation arrangements in the safe system of work for maintenance work on grouting or slurry lines, especially if removing guards.
- The fitting of anti-whip devices (whip checks) across flexible hose connections to prevent the ends flying about if they become disconnected under pressure.

TUNNELLING

- Being aware that pumped grouting and sprayed concrete operations can result in the following.
    - Injury resulting from burst hoses.
    - Injury resulting from blowout at the injection point.
    - Hearing damage and loss through exposure to excessive noise levels.
- Immediately releasing the pressure if a grouting or sprayed concrete hose becomes blocked.
- Cleaning grouting pipelines after use to prevent blockages and bursting when they are used again.
- Replacing hoses that show signs of swelling, which suggests they are damaged.
- Being aware that hoses, which are of a similar size but with different markings, are **not** interchangeable. Although hoses may appear similar, they may have one of the following attributes.
    - Have different operating capacities.
    - Be for different uses.
    - Be of different physical sizes.

TUNNELLING

26

CONTENTS
Plumbing (JIB)

What your site and employer should do for you	226
What you should do for your site and employer	226
Introduction	227
Safe systems of working	227

PLUMBING (JIB)

What your site and employer should do for you
1. Make sure that a competent person supervises the work.
2. Tell you about the hazards identified in the risk assessment.
3. Provide you with a safe system of work that is without risks to your health.
4. Explain the safe system of work to you in a method statement.
5. Give you adequate instruction and training.
6. At no time place you in danger.

What you should do for your site and employer
1. Follow the agreed safe system of work.
2. Keep to any permit systems which are in operation.
3. Not take risks with your own or anyone else's health or safety.
4. Report any aspect of your work that you feel is unsafe.

PLUMBING (JIB)

Introduction

Plumbing and gas-related work is highly specialised and potentially dangerous. All such work must be properly planned, only carried out by trained and competent contractors and be adequately supervised.

No-one should attempt to work on any gas pipework or equipment unless they are a Gas Safe registered engineer.

 Poorly thought out or executed gas-related work activities can kill.

Safe systems of working

Safe working must include attention to the following.

- Looking after equipment and hand tools to make sure they are safe to use, including carrying out simple repairs (such as replacing a split file handle) where this is practical.
- Being aware that cutting large diameter metal pipes will leave extremely sharp edges.
- Leaving all places of work in a safe condition if they have to be left unoccupied, particularly if working on gas systems in domestic premises where the occupier will not be risk-aware.
- Making sure there is safe access to any place of work and that the workplace itself is safe to occupy (for example, using a ladder or stepladder if accessing the loft space of a domestic premises and making sure that there is safe access over the joists).
- Carrying out the transportation of people, equipment and materials in a safe and responsible manner, such as in the following examples.
  - Only carrying people in a van if it is fitted with factory-fitted seats and seat belts.
  - Carrying long lengths of tubing in a suitable pipe-rack attached to the roof of a van.
  - Always wearing a seat belt, when one is provided, if operating construction plant.
- Not entering an excavation if the sides are not supported or if the supports show signs of collapse.

 Also refer to the 'Common elements' section of Chapter F28 Heating, ventilation, air conditioning and refrigeration (HVACR).

PLUMBING (JIB)

 For further guidance refer to:

Joint Industry Board PMES England & Wales
Lovell House, Sandpiper Court
Phoenix Business Park
Eaton Socon
St Neots
Cambridge
PE19 8EP

Tel: 01480 476925

www.jib-pmes.org

CONTENTS

Heating, ventilation, air conditioning and refrigeration (HVACR)

What your site and employer should do for you	230
What you should do for your site and employer	230
Introduction	231
Common HVACR elements	231
Domestic heating and plumbing services (HAPS)	242
Pipefitting and welding (industrial and commercial) (PFW)	242
Ductwork (DUCT)	243
Refrigeration and air conditioning (RAAC)	244
Services and facilities maintenance (SAF)	245

HEATING, VENTILATION, AIR CONDITIONING AND REFRIGERATION (HVACR)

What your site and employer should do for you
1. Make sure that a competent person will supervise the work.
2. Tell you about the significant hazards identified in the risk assessment.
3. Provide you with a safe system of work that is without risks to your health.
4. Make sure you understand the safe system of work and the method statements.
5. Give you adequate instruction, training and supervision.
6. At no time expect you to work unsafely or work in a place of danger.

What you should do for your site and employer
1. Follow the agreed safe system of work.
2. Keep to any permit systems that are in operation.
3. Not take risks with your own or anyone else's health or safety.
4. Report any aspect of your work that you feel is unsafe.

HEATING, VENTILATION, AIR CONDITIONING AND REFRIGERATION (HVACR)

Introduction

The information in this common heating, ventilation, air conditioning and refrigeration (HVACR) chapter should be read by people from the following trades.

- Domestic heating and plumbing services (HAPS).
- Pipefitting and welding (industrial and commercial) (PFW).
- Ductwork (DUCT).
- Refrigeration and air conditioning (RAAC).
- Services and facilities maintenance (SAF).

You should also read the section later in the chapter that refers to your particular trade.

Common HVACR elements

Competency

Competency is essential if work is to be carried out in a way which is safe and, so far as is reasonably practicable, free of risks to health, safety and the environment. Competency can be defined as a combination of training, knowledge, attitude, experience and skills.

Competency is a 'two-way street'.

- No-one should ask any other person to carry out a job unless they know that the other person is competent, adequately resourced and able to carry it out in a way that is safe and without risks to health and the environment.

- No-one should accept any job unless they know that they are competent and have the resources to carry it out in a way that is safe and without risks to health and the environment.

Examples of good practice are listed below.

- All work must be properly planned in advance, with the risks assessed and eliminated or controlled.

- Only Gas Safe registered engineers are authorised to work on fuel gas pipework or components.

- Dangerous fuel or other gas fittings that could cause a death or major injury must be reported to the Health and Safety Executive (HSE).

- The pressure testing of pipework and vessels must only be carried out by someone who is trained, competent and authorised.

- All power tools must be used in a safe and responsible way.

- Only people who are trained, competent and authorised and work for an F-gas registered company are allowed to install, service or maintain systems that contain or are designed to contain refrigerant gases.

- Anyone who has a concern about their own or anyone else's health or safety must report it to someone in authority on the site.

HEATING, VENTILATION, AIR CONDITIONING AND REFRIGERATION (HVACR)

Hot work

Hot work presents the obvious risk of fire. Many serious fires have occurred because the risks were not properly managed. All hot works must be planned for and managed correctly. One way of managing the risk of fire resulting from hot works is the implementation and policing of a hot-works permit system. Anyone planning to carry out hot works that involve the use of a blow torch or hot air gun should make sure of the following.

- Seek authority and agreement from site management.
- Obtain approval from your supervisor or site manager.
- Make sure that an extinguisher of the correct type is available in the immediate area.
- Remove lagging from pipework for at least 1 m either side of where work will be carried out on lagged pipes.
- When using a blowtorch near to any combustible material (such as timber), place a mat of a non-combustible material over the combustible material to prevent it catching fire.
- Stop any work that needs a hot-work permit, in line with the time requirements defined in the hot-work permit (this is usually at least one hour before leaving the job). Inspect the area before leaving.
- Notify your supervisor or site manager when the hot works are complete.

Anyone planning hot works must implement a hot-work permit scheme and make sure it is followed.

Checking, installing, testing and commissioning installations

- Plant and equipment must be installed, tested and commissioned in such a way that it is safe to use.
- It is essential that the unauthorised use of plant and equipment being installed is prevented. If necessary, lock off switches and valves until such time as the equipment has been fully commissioned.

Confined spaces and risers

Confined space working is recognised as being particularly hazardous.

- No-one should enter a confined space unless:
  - a risk assessment has been carried out
  - a method statement has been prepared
  - if required by the site or the employing organisation's rules and procedures, a permit to work has been issued.
- A Gas Safe certificate may be required prior to entry into a confined space.

Danger Confined space
No unauthorised entry
Permit to work must be obtained

Confined space signage

HEATING, VENTILATION, AIR CONDITIONING AND REFRIGERATION (HVACR)

- Other ways of doing the job must always be investigated before entry is made.
- The use of hazardous substances, which can be breathed in, must be closely controlled in confined spaces.
- Carrying out confined space working in an unsafe way has caused many deaths. Many of the victims were people who were trying to perform rescues for which they were not trained or equipped.
- If a naturally occurring toxic or flammable gas (such as hydrogen sulphide or methane) is detected, the space must be evacuated immediately.
- Potential failure of the lighting in a confined space must be controlled by issuing workers with intrinsically safe (will not cause spark or ignition) torches, as a minimum.
- If oxyacetylene equipment is used in a confined space, the two main safety considerations are the following.
    1. Unburnt oxygen, causing an oxygen enriched atmosphere.
    2. A flammable gas leak.

 A worker died after inhaling toxic fumes while carrying out restoration work in a bathroom.

Simply providing forced ventilation may not be sufficient to create or provide respirable (safe to breathe) air.

Electrical safety

Electricity can kill. It cannot be detected by any of the senses except touch. Working close to or near exposed live electrical circuits can result in a fatal shock. Here are some examples of good practice in electrical safety.

- Defective electrical equipment, including hand tools, must be taken out of service immediately and a procedure put in place to make sure they cannot be used.
- Extension leads must be run out in a safe way so that they do not overheat and are not a tripping hazard. If possible run them above head height or along the join of the walls and the floor.
- Electrical distribution circuits must only be installed by competent and authorised electrical contractors.
- If the supply system does not meet your needs, tell someone and then stop work until an authorised supply has been installed.
- All portable 230 volt and 110 volt equipment must be periodically tested for electrical safety: portable appliance test, commonly known as *PAT testing*.
- Battery-powered tools do not need testing although mains-powered battery chargers do.
- Temporary continuity bonding must be installed before breaking into metal pipework to provide a continuous earth for the installation throughout the duration of the work.

HEATING, VENTILATION, AIR CONDITIONING AND REFRIGERATION (HVACR)

- Electrically powered hand tools should not be used to carry out work outside in wet weather.
- Testing for hidden cables within the structure of a wall should be carried out, using a cable tracer, before disturbing the fabric of the wall.
- Carry out the following before working on electrically powered equipment.
  - Make sure the equipment is switched off.
  - Isolate the supply at the main board.
  - Lock out and tag the circuit at the main supply board.
  - Prove the circuit dead with a GS38 proving unit and test instrument.

Electrical lock-off devices

- If work has to be carried out on electrical equipment and the main isolator does not have a lock out device, the person(s) doing the job should carry out the following.
  - Identify if the circuit to be worked on is fed from a single or multiple supply.
  - Using appropriate methods isolate the circuit from all sources of supply.
  - Secure the means of isolation.
  - Display clear warning signs at the isolators that the equipment is being worked on and must not be energised.
- If working on or near to live exposed conductors, a safe system of work (including a permit to work) should be in place in accordance with the requirements of the Electricity at Work Regulations.
- The use of 110 volt power tools is preferred on site because they are safer; transformers are used to reduce the 230 volt supply to 110 volts.
- Electrical power cables are often colour coded – 110 volt (yellow), 230 volt (blue) and 400 volt (red).
- Do not assemble a mobile access tower near to overhead electrical cables until it is confirmed that the cables are dead.
- No-one, other than competent persons engaged on authorised work, must work on or near to exposed electrical conductors unless these are confirmed to be dead. Damaged cables must be isolated and replaced.
- The padlock to an electrical lock-out guard can be fitted by anyone working on the equipment. However, only competent and authorised persons should work on electrical distribution systems.
- If the mains isolator for a piece of equipment is found switched off upon arrival on site, work must **not** start until the person in control of the premises has been spoken to and the item is proven to be disconnected from all electrical supplies. Secured disconnection (isolation) should be achieved.

HEATING, VENTILATION, AIR CONDITIONING AND REFRIGERATION (HVACR)

- Adequate task lighting must be provided, where there is not enough natural light to allow any job to be carried out safely.

Emergency situations

If you are the first person on the scene following an accident to another worker, it is important that you know what to do. Carry out the following if you find anyone who is injured.

- Make sure that you are not in any danger.
- Stay with the victim, keep them still and send someone else to find a first aider.

It is important that everyone on site knows what to do in an emergency situation. A serious emergency may result in the evacuation of the whole site.

- If a natural gas leak in an enclosed area is reported, the area should be ventilated and the gas emergency service should be contacted.
- The following steps must be taken if a refrigerant leak is reported in an enclosed area.
    - The area must be ventilated.
    - All naked flames must be extinguished.
    - It must be established whether or not it is safe to enter the area before anyone tries to do so.

Health risks

Health risks are often overlooked because the symptoms are not immediately obvious. Examples of health risks, together with actions to take to make sure that the risks are managed, are listed below.

- Asbestos is likely to be found in any building built before the year 2000. Anyone who is likely to disturb the fabric of a building must be aware of the following.
    - Asbestos could be present.
    - Asbestos can be found in many places.
    - Asbestos-containing materials (ACMs) and products include insulation boards around radiators, gaskets and seals in joints, and rope seals in a boiler.
    - If the presence of asbestos is suspected, work must stop immediately and the situation must be reported to a supervisor or manager.

    Note: the asbestos register must be checked before the work starts.

- Anyone who is likely to disturb asbestos should be trained, and the training must be suitable for the work being undertaken and must be refreshed annually.

- Anyone suffering from a headache or sickness whilst using a solvent-based product (such as an adhesive) should be taken into fresh air. They will also need first-aid treatment, and the incident should be reported.

- Discarded items of drug-using equipment (for example, hypodermic needles) must be safely removed by wearing gloves and using grips if practical. The supervisor or manager must be told and the incident should be reported internally.

HEATING, VENTILATION, AIR CONDITIONING AND REFRIGERATION (HVACR)

- Manual handling must be carried out in a way that avoids injury.
    - Generally assessing the task as a whole before trying to lift items.
    - Operatives telling their supervisor and asking for help to move any load that they know is too heavy for them to move without help.
    - Using a suitable manual handling aid (such as a trolley) to move loads, particularly over long distances.
- Repeatedly bending copper tube over the knees using an internal spring could result in long-term damage to the knees.
- Noise assessments must be carried out by a competent person to assess if there is a danger of noise-induced hearing loss, or nuisance to neighbouring residents.
- When carrying out solvent welding on plastic ductwork it is essential that the area remains well ventilated and that the appropriate respiratory protective equipment (RPE) is worn.
- You can avoid risks to your health from working with lead by carrying out the following.
    - Not smoking or eating whilst bossing or otherwise handling lead.
    - Preventing lead from getting into the bloodstream by washing your hands after handling it.
    - Using appropriate extraction systems and RPE when soldering or welding lead.
- The potentially fatal risk from legionella can be controlled by being aware of the following.
    - The ideal temperature range for the bacteria to breed is between 20°C and 45°C.
    - Breeding grounds for legionella include slow-moving or stationary water supplies (such as infrequently used shower heads or 'dead legs') that are within the above temperature range.
    - The bacteria are spread to humans through breathing in water droplets in the form of fine mists and sprays.
    - Suitable RPE designed to catch mist and water droplets with a protection factor of 40 must be worn when breaking into the system, if exposure to sprays or mists cannot be prevented.
    - Symptoms of legionella are headache, muscle pain, chills and high temperature. This will lead to coughing, shortness of breath, chest pains, vomiting, diarrhoea and confusion.
    - If a case of legionella is suspected, the HSE must be informed immediately.

HEATING, VENTILATION, AIR CONDITIONING AND REFRIGERATION (HVACR)

LPG and other gases

Liquefied petroleum gas (LPG) is a flammable and explosive gas that is often found on construction sites. Like many other compressed gases it is dangerous if not handled and stored properly. Some safety precautions are listed below.

- LPG cylinders that supply site cabins must be kept outside of the cabin.
- LPG and acetylene cylinders must be stored in the open air to prevent the build-up of leaking gas.
- If LPG cylinders are transported and/or stored overnight in a van, there must be a low-level drainage route to the outside to protect against the build up of leaking LPG.
- Where LPG cylinders are being transported in a van, they must be carried in a purpose-built container inside the vehicle and have an LPG sign on the outside of the vehicle.
- Anyone who has to transport more than 5 kg of LPG in a covered van must be trained and competent in the hazards relating to the gases.
- It is important to know the difference between equipment that is used on propane and butane because propane equipment works on a higher pressure.
- LPG is heavier than air and leaking gas will collect at the lowest point it can reach (such as basements, drains or gulleys).
- Except for cylinders designed for use on gas-powered forklift trucks, store all LPG cylinders in an upright position to prevent the liquid gas being drawn from the cylinder and creating a hazardous situation.
- If a leak is suspected in any part of an LPG system efforts to trace it must only be made using a proper leak detection fluid.
- Oxygen cylinders must be handled with great care due to the high pressure to which they are filled.
- Make sure of the following when using oxyacetylene equipment.
  - The cylinders must be secured in an upright position.
  - The cylinders, hoses and flashback arrestors must be in good condition.
  - The area must be well ventilated and clear of any obstructions.
  - The cylinders must be stored separate from other types of gas cylinder in a special compound.
  - It should not be used for jointing copper tube using capillary soldered fittings.
- Acetylene cylinders must also be stored away from other flammable and combustion supporting gases, outside in a special storage compound.

HEATING, VENTILATION, AIR CONDITIONING AND REFRIGERATION (HVACR)

- It is important to be able to identify the content of gas cylinders by their colour. For this reason they are colour coded.
  - Propane cylinders are red or orange.
  - Butane cylinders are blue.
  - Acetylene cylinders are maroon.
- Where welding is being carried out, screens must be provided to protect others from welding flash.

Lifting operations

Lifting operations have been the cause of many accidents because they were not properly planned, carried out or defective equipment was used. Everyone involved in lifting operations must be aware of the following.

- The sequence of operations in any lifting activity must be planned and written down in advance in a method statement (called a lift plan).
- The rated capacity (SWL) of any piece of lifting equipment or accessory must never be exceeded.
- Every piece of lifting equipment and each lifting accessory must be marked with its rated capacity (SWL).
- If any item of lifting equipment is found to be defective the equipment must not be used, the item should be placed in quarantine (taken out of service) and the problem reported to a supervisor or manager.
- All lifting equipment must be stable and secure when being used.
- Items that are lifted into a place where they will be installed must not be released from the lifting equipment until securely fixed in place.

All lifting equipment must be thoroughly examined periodically by a competent person.

Lifting accessories (such as chains, strops and lifting eyes) and lifting equipment used to lift or lower people must be thoroughly examined every six months. All other lifting equipment must be thoroughly examined every 12 months.

Personal protective equipment

Wearing the correct personal protective equipment (PPE) can prevent exposure to harmful substances. The wrong type of PPE is unlikely to provide protection from a specified hazard. PPE should always be considered as a last resort (for example, when a risk remains that cannot be controlled by other means higher up the hierarchy of preventative and protective measures). Wear the following types of PPE in the situations described.

HEATING, VENTILATION, AIR CONDITIONING AND REFRIGERATION (HVACR)

- The correct type of impact-resistant eye protection when drilling, cutting or grinding any material that could produce flying debris.
- Thermally protective gloves to avoid direct skin contact with pipe-freezing equipment. Always read the control of substances hazardous to health (COSHH) assessment for the product.
- A safety helmet, protective footwear, hearing protection, suitable RPE and eye protection when using a hammer drill.
- Ear defenders if working in a noisy environment or doing any job in which high levels of noise are produced.
- Suitable eye protection if carrying out oxyacetylene welding.
- Safety gloves, suitable RPE and eye protection, if having to handle fibreglass roof insulation.
- RPE when using or creating hazardous substances that can be breathed in.
- Cut-resistant gloves when handling anything with sharp edges.

 PPE should only be considered as a last resort.

Arc welding masks

Safety gloves

Safe systems of work

The importance of adopting safe systems of work, taken from a risk assessment, which includes a method statement or similar document, cannot be overstated. Some examples are shown below.

- The correct hand tools must be used for any job (such as using a hammer and bolster chisel for taking up floorboards).
- The transportation of tools and materials must be carried out in a safe and responsible way, as in the examples below.
    - Carrying long lengths of tubing in a pipe-rack attached to the roof of a van.
    - Fixing a ladder to the roof rack of a van using proper ladder-clamps.

HEATING, VENTILATION, AIR CONDITIONING AND REFRIGERATION (HVACR)

- The exhaust fumes from vehicle exhausts are toxic. If internal combustion engine-driven plant has to be run inside a building, the exhaust fumes must be extracted to a safe position outside the building.
- Prevent contact with moving parts of machinery or rotating materials by enclosing them with guards or barriers.
- Be aware of sharp edges when cutting through sheet material or pipes.
- Keep unauthorised people out of the area when pressure testing pipework or vessels.
- Prevent the entanglement of clothing when using any rotating machine.
- A permit to work system should be introduced for all high-risk work activities.
- Before starting work on any piece of equipment it is essential to check the operation and maintenance manual for the equipment.
- To prevent unauthorised access to plant and switchgear rooms the doors must always be kept locked and only competent and authorised persons should be able to access them.
- Where possible lone working should be avoided. A lone worker must register their presence with their site management or site representative before starting work, and a system of periodically checking on the lone worker must be in place.
- Spilled liquids (for example, oils and greases) can create a slip and/or trip hazard. Refuelling and decanting should be carried out in an approved place using nozzles, funnels and on a spill mat if possible. If a spill does occur, contain the spill using a spill kit, tell your supervisor and keep people out of the area until the spill is cleaned up.

 Do not take short cuts. Always follow the safe system of work and use the correct tool for the job.

Working at height

Examples of good practice for safe working at height are listed below.

- A risk assessment must be carried out before any work starts.
- If a ladder is used, it should be secured to the structure against which it is resting to prevent it from slipping.
- When using a stepladder, the restraining mechanism must be fully extended.
- Ladders and stepladders must only be used in the following situations.
  - If a risk assessment shows they are suitable for the job.
  - They are in good condition.
  - For light, short-term work (lasting no more than 30 minutes) that does not involve stretching or reaching.
  - Where the work does not involve twisting the torso. If you cannot work in a forward facing position, a ladder is the wrong platform.

HEATING, VENTILATION, AIR CONDITIONING AND REFRIGERATION (HVACR)

- When a mobile access tower is used, occupy only one working platform at any one time.

- When deciding the maximum height to which a mobile access tower may be built, refer to the manufacturer's instructions.

- Immediately close the hatch of a mobile access tower working platform after gaining access to the platform.

- Before moving a mobile access tower, remove all people, tools and equipment from the working platform.

- The users of mobile access towers must be made aware of overhead live electric cables, and make sure that these are disconnected or isolated, before putting up the tower.

Proprietary access system with ladder access, working platform, guard-rails and hoist

- Edge protection must be fitted to working platforms to prevent the fall of a person or materials.

- A stable working platform (such as a mobile access tower) should be used where the job will take longer than 30 minutes, there is a good floor surface to ensure stability of the scaffold and 'heavier' type work is to be carried out at height.

- Ladders used on site must be labelled 'Class 1' Industrial or 'EN 131' Trade and Industrial – heavy duty and industrial use (for professional users).

- Wooden ladders that have been painted must not be used because the paint can hide defects.

- Any opening (such as shafts, pits, service ducts and large floor voids) must be fitted with double guard-rails and toe-boards or a secure cover to prevent someone falling through them.

- If working at height to dismantle lengths of soil pipe, the job can be safely carried out by working in pairs and breaking the length at the collar to remove complete sections.

- Flue liners must be installed in a safe way by working in pairs at roof level and using a safe method of access (such as a chimney scaffold).

- As far as practical, using a ladder to access ceiling voids or soffits containing a large number of services should be restricted. Other means of access may be needed.

- Never use utility service pipes or cables as a makeshift working platform or a means of access.

- Objects must not be thrown or dropped from height and must be prevented from falling.

HEATING, VENTILATION, AIR CONDITIONING AND REFRIGERATION (HVACR)

Domestic heating and plumbing services (HAPS)

Installing or maintaining heating and plumbing services can present the risk of injury or ill health. Some examples of good practice are listed below.

- Temporary continuity bonding should be carried out before removing and replacing sections of metallic pipework to provide a continuous earth for the installation.
- When welding is being carried out, a screen should be provided to protect others from welding flash.

> If a job involves work below a ground-level suspended timber floor, you should first check or ask if the work could be performed from above it, rather than entering a potentially confined space with restricted working room.

Pipefitting and welding (industrial and commercial) (PFW)

Pipefitting and welding can present the risk of injury or ill health to you and others. Some examples of good practice are listed below.

- When using a pipe threading machine, a safety barrier should be erected around the whole length of the pipe and you should make sure that your clothing cannot get caught on rotating parts of the machine. In accordance with the requirements of the Provision and Use of Work Equipment Regulations (PUWER) all dangerous parts of machines shall, so far as is reasonably practicable, be fenced or guarded.
- When a new piece of plant has been installed but has not been commissioned, it should be left with all valves and switches locked off.
- Before using oxyacetylene equipment it is essential to check the following.
  - The cylinders, hoses and flashback arresters are in good condition.
  - The area is well ventilated and clear of any obstructions.
- Acetylene cylinders (maroon in colour) should be stored outside in a special storage compound when not in use.
- When using oxyacetylene brazing equipment, the cylinders should be stood upright and secured, preferably on a purpose-built trolley.
- When using pipe-freezing equipment to isolate the damaged section of pipe, you should read the COSHH assessment and wear appropriate PPE and RPE.

When carrying out pressure testing of pipework or vessels, exclusion zones must be in place and access only given to those involved in the testing.

HEATING, VENTILATION, AIR CONDITIONING AND REFRIGERATION (HVACR)

Ductwork (DUCT)

Installing or maintaining ductwork can present the risk of injury or ill health. Some examples of good practice are listed below.

- Before painting the external surface of any ductwork, you should read the COSHH assessment for the paint.
- Fume extraction must be provided when welding or carrying out other hot work on galvanised ductwork.
- Before using a solvent based adhesive on ductwork you must:
  - read the (manufacturer's) safety data sheet and the COSHH assessment
  - establish a safe working area
  - comply with the COSHH assessment (for example, the creation of confined spaces conditions, ventilation, PPE, smoking, and sources of ignition).
- If, before fitting, a defect is noticed in any component of a system that is being installed, the item must not be fitted and you must tell your supervisor or manager.
- Before using a cleaning agent or biocide on a ductwork system:
  - read the (manufacturer's) safety data sheet and the COSHH assessment
  - speak to the building occupier and establish a safe working area and conditions that are safe for yourself, occupants of the building and others
  - produce a method statement for the work in compliance with the COSHH assessment (for example, the creation of confined spaces conditions, ventilation, PPE, smoking and sources of ignition).
- Before cleaning a system in an industrial laboratory or other premises where harmful particulates might be encountered:
  - the system must be inspected
  - samples must be collected from the system
  - a job-specific risk assessment and method statement must be prepared.
- If it is necessary for a person to enter ductwork, two items that must be considered are the:
  - dangers of working in confined spaces
  - strength of the ductwork and its supports.
- Before working on a kitchen extraction system, the nature of the cooking deposits in the system should be established.
- Aluminium ductwork that has been pre-insulated with fibreglass must only be cut using tin snips, with the people involved in the job wearing suitable PPE and RPE.
- When jointing plastic-coated metal ductwork, welding presents far greater health risks than, for example, riveting, taping or using nuts and bolts.

HEATING, VENTILATION, AIR CONDITIONING AND REFRIGERATION (HVACR)

 If dismantling extract ductwork, it is essential to find out what it may be contaminated with before starting work.

Refrigeration and air conditioning (RAAC)

Refrigerants and other gases

Most gases have the potential to cause harm because they are explosive, stored at high pressure, are at low temperatures or are harmful to the environment. Examples of good practice and the unsafe properties of gases are shown below.

- When not in use, refrigerant cylinders must be stored in a special, locked storage compound in the open air.
- Refrigerant gases are heavier than air and if released into an enclosed space will sink to the lowest place they can seep into.
- If transported in a van, cylinders of refrigerant gases must be carried in a purpose-built container inside the vehicle.
- When handling refrigerant gases, eye protection, overalls, thermal resistant gloves and safety boots must be worn.
- Flashback arrestors must be fitted between the pipes and gauges of oxypropane brazing equipment.
- Make sure you check the following when pressure testing using nitrogen.
    - The gauges can take the necessary pressure.
    - The nitrogen cylinder is secured in an upright position.
- Oxygen must never be used for pressure testing because it could react with the oil in the compressor causing an explosion, possibly resulting in serious injury or death.

Safe systems of work

- When a refrigerant leak has been reported in a closed area, it must be established that the area is safe before anyone tries to enter.
- Before entering a cold-room it must be established that the exit door is fitted with an internal handle.

HEATING, VENTILATION, AIR CONDITIONING AND REFRIGERATION (HVACR)

Services and facilities maintenance (SAF)

Checking, installing, testing and commissioning installations

Where new plant or equipment is being installed it must be done in a way that avoids injury to the people doing the job or to anyone else. This must include making sure that the plant or equipment cannot be operated in an unauthorised way or by unauthorised persons before it has been commissioned and handed over.

- The health and safety file for any building is a useful source of information on the safe way of maintaining the systems within it.

- The temperature of the hot water at the tap furthest from the boiler should be at least 50°C within one minute of starting to run it.

- Two minutes after running a cold water tap its water temperature must be below 20°C.

- Where there is a cooling tower on site, it must be registered with the Local Authority and have a formal log book that is kept up-to-date.

- On cooling tower systems, the water must be chemically treated.

- Make sure the following actions are carried out when replacing the filters in an air-conditioning system.
    - A job-specific risk assessment and method statement must be prepared and followed.
    - The person(s) doing the job must wear suitable overalls and a respirator.

- After a gas boiler has been serviced it must be checked for the following.
    - Flueing.
    - Ventilation.
    - Gas rate.
    - Safe functioning.

- Before adding an inhibitor to a heating system, the COSHH assessment for the product must be read and understood.

Pressure systems

The following are examples of designated pressure systems that need a written scheme of examination and a safe system of work under the Pressure Systems Safety Regulations 2000.

- Steam sterilising autoclave and associated pipework and protective devices.
- Steam boiler and associated pipework and protective devices.
- Gas-loaded hydraulic accumulator, if forming part of a pressure system.
- Portable hot water/steam-cleaning unit fitted with a pressure vessel.
- Vapour compression refrigeration system where the installed power exceeds 25 kW.
- Compressors and systems where the total volume of compressed gas in the system is in excess of 250 bar/litres.
- Fixed LPG storage system supplying fuel for heating in a workplace.

HEATING, VENTILATION, AIR CONDITIONING AND REFRIGERATION (HVACR)

CONTENTS
Further information

Training record — 248

FURTHER INFORMATION

Training record

Name of company _____

Name of employee _____

Name of supervisor _____

Instructions to supervisor

The employee and supervisor should sign each area of training listed below as it is completed. The manager responsible should endorse the record and ensure that a copy is retained on file.

	Date completed	Employee's signature	Supervisor's signature
1 General responsibilities			
2 Accident reporting and recording			
3 Health and welfare			
4 First aid and emergency procedures			
5 Personal protective equipment			
6 Asbestos			
7 Dust and fumes (Respiratory hazards)			
8 Noise and vibration			
9 Hazardous substances			

FURTHER INFORMATION

	Date completed	Employee's signature	Supervisor's signature
10 Manual handling			
11 Safety signs			
12 Fire prevention and control			
13 Electrical safety			
14 Work equipment and hand-held tools			
15 Mobile work equipment			
16 Lifting operations and equipment			
17 Working at height			
18 Excavations			
19 Underground and overhead services			
20 Confined spaces			
21 Environmental awareness and waste control			
22 Demolition			
23 Highway works			
24 Specialist work at height			
25 Lifts and escalators			
26 Tunnelling			
27 Plumbing (JIB)			
28 Heating, ventilation, air conditioning and refrigeration (HVACR)			